BIODYNAMIC
WINE GROWING

BIODYNAMIC WINE GROWING

Understanding the Vine and Its Rhythms

JEAN-MICHEL FLORIN

 Floris
Books

Translated by Bernard Jarman

First published in German under the title
Biologisch-dynamischer Weinbau:
Neue Wege zur Regeneration der Rebenkultur
by Verlag Am Goetheanum in 2017
First published in English in 2021 by Floris Books

 Also available as an eBook

British Library CIP available
ISBN 978-178250-669-0
Printed in Poland through Hussar

 Floris Books supports sustainable forest management
by printing this book on materials made from wood that
comes from responsible sources and reclaimed material

CONTENTS

INTRODUCTION

Like wheat, vines have accompanied human development since ancient times. Compared to all other areas of conventional agriculture, viticulture today uses the largest number of chemical treatments in order to 'protect' its vines. This situation came about in a dramatic way as a result of the grape louse depredations more than 150 years ago. The vine is even referred to as a 'museum of plant pathologies'.

Rather than seeking to solve each problem on its own and thereby merely address symptoms, we need to recognise that the vine is severely weakened: it has become sick and ever more prone to disease and pest attack – thanks largely to the 'modernisation' of vine cultivation.

How did we get to this point? What is the current situation? And what possibilities are there for the future regeneration of the vine and viticulture?

More and more wine growers are waking up to this situation and looking to change their approach in a fundamental way and understand the vine better. Since the 1980s wine growers have been converting to the biodynamic approach in increasing numbers. They are often amazed at the positive response of the vine and indeed the whole vineyard: improved soil, healthier plants and above all better grape and wine quality. In many countries (Germany, Italy, France, Spain, USA) biodynamic viticulture is becoming the public face and ambassador for biodynamic agriculture. Some of the world's best wines are now biodynamic.

Over the decades growers have had many positive experiences with biodynamic viticulture. But of course, many new questions have arisen about the future of the vine itself. This book offers an interim summary of the wide range of experiences with biodynamic viticulture in order to develop a vision for the future of vine cultivation.

This book follows three steps. The first step (Parts 1 and 2) offers a holistic, phenomenological observation of the vine. If we want to understand the current situation of the vine we need to try and understand its inner nature. The more intimate understanding of the vine made possible by these observation techniques enables us to determine whether the practical measures applied are appropriate to the vine or whether they weaken the plant. Knowledge of the vine's cultural history, and the way it has changed over the last 150 years, is also part of this.

In the second step (Parts 3 and 4), consideration will be given to the conditions required for healthy vine cultivation – how should the vine of today be cared for to ensure it remains healthy? In this section various aspects of biodynamic practice will be described, drawing on concrete examples.

But in the long term none of this will be sufficient. A third step is needed – how can weakened vines be fundamentally regenerated? And what is the future mission of this plant? The third section of the book (Part 5) will explore the possible direction of travel for practice-based research.

May this book give its readers inspiration for the future of the vine.

PART 1

A Goethean Approach

1

THE ARCHETYPAL PLANT

Jean-Michel Florin

The qualitative-phenomenological approach developed by Johann Wolfgang von Goethe (1749–1832) was deepened and taken further by Rudolf Steiner. Through it, the plant can be understood in its inner nature as a living entity. The approach begins by simply using our senses (morphology, taste, texture, etc.) to observe the plant. According to Goethe: 'The unique quality of this approach is that the information for evaluation does not come from oneself but from the things that are being observed themselves.'[1]

In order to take a new and fresh look at sense-perceptible phenomena in a Goetheanistic way, everything which is already known and all preconceived ideas must be laid to one side. A Goethean approach means that sense perceptions need to become more refined. Goethe invites us to regain trust in our senses:

> In so far as the senses are healthy, the human being itself is the greatest and most precise physical apparatus of all; and what is most harmful about modern physics, is the fact that human beings have to separate themselves from the experiments and allow only the results revealed by artificial instruments to describe nature.[2]

Every vintner who smells, tastes and contemplates the wine to assess its maturity and particular quality, uses the sense of smell and taste in a very refined way. The more we can train our senses in this way, the more diverse will be the range of phenomena we can gather about the plant, its environment or its life cycle. It is very helpful to ask the question 'how?' By asking 'how?' we come into a relationship with the plant's inner nature which then comes to expression in the way it manifests itself.

In order to take these newly discovered phenomena seriously we need to develop a quality of wonder. Hermann Hesse, the famous writer who made a study of Goethe, expressed it as follows:

> With wonder it begins and with wonder it ends and yet it is no futile journey. Whether I stand in awe before some moss, a crystal, a flower, a golden beetle, or a cloudy sky, the calmly rolling swell of the ocean, the crystalline network of veins on a butterfly's wing, its clear-cut shape and colourful edges, its diversity of form and patterning and the endlessly delicate and magically changing coloration – whenever I cast my eye upon or experience a piece of nature with one of my other senses, when I feel drawn to and enchanted by it and for a moment open myself up to its presence and what it reveals, in that same instance I forget the entire world of human greed and blind pursuit of objects, and instead of planning or giving orders, instead of procuring or exploiting, fighting or organising, I do nothing in that moment but like Goethe, am filled with 'wonder'.[3]

The question 'for what?' or 'why?' leads to a causal explanation and a plant becomes reduced to its functionality. A living being however, unlike a machine, is not solely defined by its function. If for example we respond to the question 'why does the vine have tendrils?' with 'so that the plant can climb up more easily', this reductionist answer precludes any further considerations regarding the plant's nature. If, however, we ask ourselves 'how does the vine form its tendrils?' we immediately make connections with other phenomena or aspects such as the vine's capacity to be open towards its surroundings, to be sensitive throughout its life.

In a similar way the question 'how does the plant develop?' leads us into its process of growth in time. Then we are taking seriously its existence as a living being instead of merely seeing it as a finished object.

Expanding consciousness

The following considerations will allow us to know the plant (the vine) ever more intimately. It will become clear in doing so that our own consciousness has to expand or change. An initial exercise will help make this change a concrete experience. The following sketches show the various growth stages of a developing vine leaf. These six drawings can be looked at in two different ways.

They can be viewed as different individual steps, with each one given a name or number. This would be a point-focused or non-continuous way of looking at it. Does such an approach, however, do justice to the reality of the process? Has the leaf not followed a continuous process of growth?

We can also attempt to follow the process of moving from one stage to the next by inwardly transforming one leaf form into another. Such an approach, requiring a different kind of thinking to the analysis of the first example, is far closer to the reality of a continuous process. We need to slip into the process of leaf formation, in other words we need to follow and recreate the process in ourselves.

Figure 1. Stages of a growing vine leaf (Vitis vinifera).

Four steps of plant observation

Step one: factual observation

After our first impressions the plant is observed as precisely and in as great a detail as possible using all the senses (sight, smell, taste, touch, etc.). Drawing is a great help towards better and more exact observation.

Step two: capturing the time structure of the plant

It is actually quite impossible to capture the growth of a plant in its entirety and without gaps. To do so would mean having to stand in front of a plant and never fall asleep. There is thus no alternative but to observe the plant at regular intervals and then inwardly to follow the processes occurring in between. On the one hand we need to observe the present state of the plant and on the other to recall its earlier forms so that through our thinking about them, both the earlier and the present form can come to life. It is in this way possible to capture the various stages of growth: germination of the seed (or opening of the seed pod), development of the stem and leaves, formation of flower and fruit. From the moment of the plant's birth until its death we inwardly take part in a process of transformation. In the subsequent imaginative recreation of the plant an understanding for the specific time dynamic of the species can be attained.

Step three: capturing the plant's gesture

All the sensory observations as well as the growth dynamics are brought inwardly together in order to discover the plant's overarching ordering principle. This may also be referred to as the plant's 'gesture'. Being receptive to the atmosphere emanating from the plant is also a requirement. This third type of observation brings us even closer to the plant's inner nature.

Step four: becoming one with the being of the plant

The fourth level of plant observation involves trying to erase all preconceptions and inner images in order to grasp what manifests itself in the unimaginable yet specific will impulse or potential of the plant's essential nature.

This step can, for example, provide an orientation for developing a culture that accords with the species' inner nature and offer guidance regarding the appropriate cultivation of the plant or as the case may be, its healing.

These steps are intended as guidelines; they should not be taken as a hard and fast framework nor should the boundaries between them be conceived too strictly. They can, however, be a help when trying to understand the specific nature of, for example, the vine.

Plant growth: point, line and plane

Point – the seed

At the beginning of its life the plant appears in the very limited form of the dormant seed. All the life processes are resting. The seed is like a speck of dust and yet with a subtle difference. This speck of dust has the potential to produce something that grows. In the seed the plant has withdrawn from the world of space and time into a point. It has internalised itself and remains largely impervious to outer influences (earth, water, light, etc.) for as long as it is dormant. The seed is nonetheless open for the regenerative influences of the cosmic environment that relate to the species. Botanists are aware that in passing through the seed stage many viruses and plant diseases vanish and even certain pathological mutations can disappear. In comparison with the fully developed, flowering plant that expresses itself in substance and form and through which it ultimately exhausts itself, the plant in its seed stage retains its highest potency. Particular importance is attached here to the possibility of a plant regenerating itself through the seed – especially in relation to the grape vine.

Line – the axis of earth and sky

When a flowering plant starts to germinate, a root develops with extraordinary power and descends vertically into the earth. It grows actively down in line with gravity towards the centre of the earth (geotropism) and in doing so is able to penetrate severely compacted soil layers. Through its engagement with gravity the root becomes more dense – immediately behind the very sensitive root tip which is so open to the soil environment, the root becomes more and more woody. Even if in some species the original root subdivides and forms secondary roots

Figure 2. The point stage – seed.

of a first, second or third order, the tendency towards tap-root formation remains more or less intact.

Soon afterwards a shoot (stem) grows vertically up out of the seed away from the earth. It grows in the opposite direction to the root and towards the light. Botanists refer to this as 'negative geotropism'. Would it not be better to call it 'sky-tropism'? The plant is, after all, growing towards the sky, and not, as is usually assumed, towards the light. Even in a dark cellar potato shoots will grow upwards, in the opposite direction to gravity. It is in fact the case that stem growth occurs most strongly at night and in shady situations. This can be very well observed with young seedlings that have too little light. Their stems grow long and they remain in their seedling stage for longer. In the case of the grape vine it is interesting that in the wild, like many climbing plants, it will start growing in the shade, develop very long stems and thereby remain at a young stage of growth.

Thus, from the very beginning, the young plant is living between two polarities – through its roots it is linked to the earth and gravity, and through its stem to the sky.

From this spatially minute point of the seed which holds the plant's entire **potential,** the plant first of all orientates itself along the earth-sky axis. It forms a line that connects these two poles. If it were to follow its own growth principle without another impulse intervening, this upward and downward growth would never stop. It is a tendency living in the growth of all climbing plants.

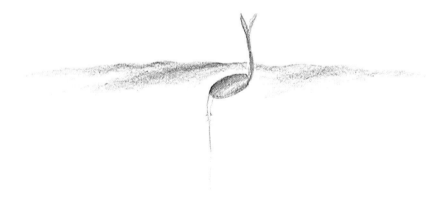

Figure 3. Forming an earth-sky axis.

Plane – opening out in the surroundings

The tip of the vertical axis – of the stem – is the main growing point that controls the vegetative growth of the whole plant. This control can be very well observed in a spruce tree – if the growing tip is cut off, numerous secondary upward-growing shoots will appear. The controlling apical dominance then no longer applies and the many side shoots compete with one another to produce a new leader.

The growing point is made up of meristem tissue out of which the leaves gradually appear. Leaves are organs of a completely new quality – they open themselves actively towards the light (upper leaf surface) and to carbon dioxide of the air (underside of leaf). This is the second stage of a plant's expansion into the environment. Up until now it has existed in space without any particular interaction – it was first necessary to find its own place – but now the plant enters into an active relationship with the elements.

After the 'point' stage of the seed and the 'line' stage of the stem we now have the 'plane' stage of the leaf.

In the case of dicotyledons two symmetrical seed leaves first appear. They seem quite primitive and reminiscent of the so-called lower (or earlier) plants. Only afterwards do the 'true' leaves gradually appear. They are organised in a spiral form around the main stem. There is an alternating damming up and extending process: a node from which a leaf develops is followed by another node from which a bud arises; in between is the inter-node, which allows the plant to stretch out into space. The nodes containing a bud generally have a powerful growth potential.

Figure 4. Sketch by Goethe of a plant's basic structure, consisting of node, inter-node, bud, leaf.

This describes the essence of the plant – stem, leaf and axle nodes – out of which a new plant can develop. This make-up offers many plants the possibility of vegetative reproduction (such as the cloning of vine rootstocks or grafting). Such a process however – unlike generative reproduction via the seed – creates a new plant using its pre-existing potential, without any renewal or regeneration occurring through the 'point stage' of the seed. In the long term, vegetative reproduction will exhaust the potential that originally existed in the seed.

Each plant can be thought of as a repeated layering of this basic structure. However, we can also observe with many plants that each new leaf takes on a new form. The basic structure of the plant is metamorphosed as it grows.

In the leaves we encounter cosmic imponderables (such as light and warmth) and terrestrial ponderables (like carbon, water and mineral salts). With the help of light the leaves transform the inorganic elements – through photosynthesis – into living substance. They breathe in carbon dioxide and breathe out oxygen, they even breathe out water.

The succession of leaves around the stem follow certain geometrical rules (phyllotaxy). While the stem exhibits a vertical tendency, there is also what Goethe describes as a 'spiralling tendency'. This reveals itself in the way the leaves order themselves round the stem. (There are interesting exceptions to the spiralling arrangement, for instance plants with decussate or fourfold arrangements.) A polarity therefore exists between the two organs of the shoot – the stem (shoot axle) which generally strives upward, and the leaves that wind around in a spiralling formation. There is likewise a polarity between the root in a root axle (with a root tip) and the root organs (for instance, root hairs).

The three types of metamorphosis

Having described the basic organs of the plant, we will now look at the growing process using the example of groundsel (*Senecio vulgaris*). We will be able to follow the dynamic of plant growth more easily if we make use of what was described earlier as living thinking.

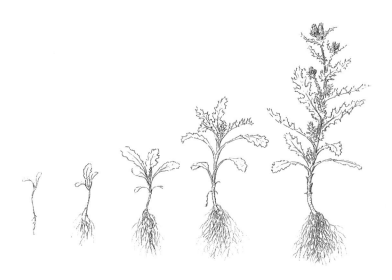

Leaf metamorphosis

In Figure 6 (opposite) we can see a series of fully developed leaves. They are laid out in their growth sequence. It is important to remember however that in nature it is rarely possible to see the entire sequence together – the first leaves have usually withered away by the time the last leaves or flowers appear. The plant never totally reveals itself at one moment. It is not a solid object but a process which develops in space and time. In order to grasp this being-of-process we must follow the various stages – germination, growth, flowering, fruiting right through to **the** production of seed – and inwardly transform them one into the other.

If we look at this leaf series and compare the different leaf forms, we can see how different form principles in the growth process succeed or permeate one another. There is first of all a process of expansion and contraction. It is important to be aware that these 'movements' or 'gestures' are not perceived by our senses but by a 'mobile thinking activity'. This basic breathing rhythm discovered by Goethe can be differentiated into a further four gestures.[4] These are described using verbs in order to make it clear that we are dealing with **active** processes and not finished stages.

🌱 In the first phase the emphasis is on the leaf stalk, which *extends* like a stem towards the periphery. This tendency can be strengthened by a rich and moist soil.

Figure 5. Development of groundsel plant Senecio vulgaris.

- In the second phase of leaf development, the leaf blade *spreads out* in all directions. This kind of form can be enhanced by shade and water.
- In a third phase the leaf *differentiates* its form, a process which is enhanced by light.
- Finally, the leaf *contracts* to a spike at the end of the stem. This tendency is enhanced by warmth and dryness.

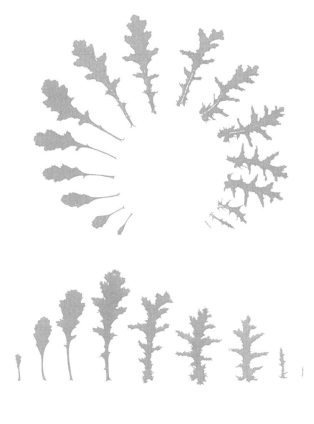

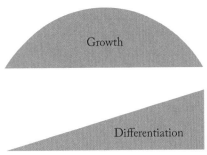

Figure 6 (above). Leaf metamorphosis of Senecio vulgaris.

Figure 7 (right). Two processes in plant development – growth and differentiation.

We can thus distinguish four motifs or formative forces. The first two (leaf, stalk and blade) are enhanced most strongly by terrestrial forces (earth and water) and have a *centrifugal* tendency. The plant fills the space with its leaves and produces a lot of mass but little form.

The last two (differentiation and contracting) are strengthened by cosmic influences (light and warmth) and show a *centripetal* tendency. The plant withdraws again from the spatial and instead of mass the focus is on form.

When looking at the leaf metamorphosis or leaf forms of various kinds of plants, we always return to these four basic motifs. Sometimes we find them in the sequence described, and sometimes they occur simultaneously, although one gesture may predominate. Shade- and moisture-loving plants tend to emphasise the first two stages, while plants that need warmth and light focus their development on the latter two stages. Just compare, for instance, the round-shaped, wide leaves of plants growing in moist woodland with the aromatic, shrubby plants in the *Labiatae* family (rosemary, thyme, savory, lavender) growing in a Mediterranean climate. We can clearly recognise in the latter, how the vegetative growth is held back by the forces of light and warmth not only in the leaves but in the entire structure of the plant.

In the vine, the first two formative forces are particularly strong, namely that of stem growth and leaf bulk. This accords with the natural environment of the wild vine, the moist, shady, humus-rich soils of the alluvial woodlands. Leaf differentiation on the other hand is more or less strongly developed depending on variety. The spike stage is found especially among the upper leaves that grow with the flowers or tendrils.

Figures 8–10. Metamorphosis in the buttercup family. Left: grassy-leaved buttercup (Ranunculus gramineus); centre: bulbous buttercup (Ranunculus bulbosus); right: lesser celandine (Ranunculus ficaria).

Flower metamorphosis

Getting to know the metamorphosis of leaves can teach us about formative forces. They are so called because they create forms. They manifest in two main ways:

- 🐚 One is a centrifugal *expanding* tendency. This expresses itself in the plant's vegetative power whereby new leaves and more substance is created through photosynthesis. This process of growth could continue endlessly if no other force were involved. The continuous growth of many twining plants demonstrates this first, formative tendency.
- 🐚 The other is a *contraction* of the leaves. The substance of the plant is becoming increasingly differentiated. This centripetal tendency, which leads towards the flower, passes through a zero point prior to forming a flower bud. If we inwardly trace the steps of leaf metamorphosis, we will notice how the leaves grow ever smaller and simpler until eventually they disappear. The stream of leaves in effect dries up. Then, following a kind of inversion (the flower bud is in essence a kind of seed on a higher level), something entirely new appears: the flower itself. This is a form that could not have been predicted. It is as though a power coming from the future gradually slows down and halts the process of growth in order to make way for something new.

The flower emerges after the plant has been devitalised. A plant with too much vegetative power, or one that has received too much fertiliser, often remains in its vegetative stage and has difficulty producing flowers. It remains immature. Every gardener will also know the opposite situation of lettuce going to seed prematurely when it is grown in poor soil, is too dry and has too much light.

The typical flowering process manifests itself through new structures that increase still further the differentiation or specialisation of the plant:

- 🐚 The stem stops growing. The plant's development does not come to a complete halt with the flower however, it continues in two new organs: stamens and ovaries, which, following fertilisation, will form the seed. The growth effectively splits itself in two. Most plants are monoecious, their flowers carry both stamens and ovaries. But some species like the stinging nettle (*Urtica dioica*), spinach (*Spinacea oleracea*) or the wild vine (*Vitis Sylvestris*) bear these organs on two separate plants. They are therefore referred to as dioecious.

- The green leaves are metamorphosed into colourful organs. The sepals and petals come together around a central point and form a kind of chalice that opens skywards. The parts of the flower are no longer directly subject to terrestrial forces (apart from a few green flowers, the flower has no possibility to photosynthesise). Instead, their relationship – especially flowers with a radial form – with the cosmic forces of light, is so strong that they open and close with the daily rhythm of the sun.

- Special quite specific substances (secondary plant substances), like nectar, pigments, scents and essential oils, arise which, being slightly flammable, demonstrate their strong connection to warmth. The plant releases them into the surroundings as scent, nectar or radiant colour and cannot use them for itself (unlike the carbohydrates formed in the green leaves).

- The various organs of the flower come together in a specific pattern and form a higher unity (some plants such as orchids bring this patterning to a high point).

- The 'typical' flower is radiant. It separates itself from the lower terrestrial region through the sepals and the crown of petals opens itself up to the realm of warmth and light. It comes into being at the very top of the stem of which it is generally the end point.

The plant's flowering pole expresses itself with quality rather than quantity. The beauty of a plant is determined not by its weight, but on the contrary by its lightness, fineness, harmony of form, the differentiation of colour, pattern, scent, etc.

Structures and substances are created in the flower which can often only be understood by observing their relationship to ensouled beings (animals and humans). This relationship can go so far in the case of orchids that they can mimic certain insects (bee orchid *Ophrys fuciflora).*

Once the fruit and seed have formed the stem generally dies back. The flowering process brings with it a devitalisation, a weakening and often even the death of the green plant. It is the death pole of the plant. The substances previously produced through photosynthesis are transformed into secondary substances.

For Goethe all the organs of the flower are transformed leaves.[5]

Fruit metamorphosis

With fruit formation the process of metamorphosis continues and the gesture here is one of contraction (the polar opposite of the leaves which expand into the surrounding space). An enrichment also occurs as it does so.

Two main processes can be observed with the formation of fruit – especially the juicy fruits which have been developed by humankind as cultivated plants:

- On the one hand, there is a vegetative process of strong cell growth in which the fruit develops rapidly. It is the green fruit which is formed. This process corresponds to vegetative growth.
- On the other hand, there is a process of ripening during which no further cell division occurs. Instead there is an increase in size, the development of flavour and a transformation of substances (increased sugar, taste, scent, colour, etc.) This process corresponds to generative growth.

Seen in this light, the juicy fruit is a kind of synthesis in that the plant simultaneously unites within itself the qualities of vegetative growth and those of flowering.

Seeds are formed inside the fruit and represent the high point of the plant's contraction and inward focus. Through the drying out of substances and inhibiting of life processes the plant is drawn back into a point.

The threefold metamorphosis in summary

The plant above the earth expresses itself in an archetypal way with three metamorphoses:

- 🐛 A *leaf metamorphosis,* in which the single leaves follow one another in space and arise over time. This is primarily a vegetative process whereby an intensive exchange with the environment brings about new living substances through photosynthesis.
- 🐛 A *flower metamorphosis,* in which the organs of the flower appear simultaneously alongside one another. A new process is active here which might be called a flowering process whereby the plant expresses its identity through very specific, differentiated forms, colours, scents, etc. It opens itself to ensouled creatures – animals (insects) and human beings.
- 🐛 A *fruit metamorphosis,* in which an interpenetration of the two (vegetative and generative) processes forms the plant organs.

At each new stage there is an enhancement of metamorphosis towards ever greater integration. We are speaking here of processes and not functions or purpose. These processes manifest differently in each plant and do not always limit themselves to the typical plant organs. Rosemary for example has fine, dry leaves that produce essential oils. In this case the flowering process works down into the leaves and transforms them. The opposite occurs in plants like the stinking hellebore (*Helleborus foetidus)* which produces green petals. Here the vegetative process rises up into the flower. In this way we see the plant as a living organism and not simply a mechanism with a number of functions.

The root

As already indicated above, there is a polarity in the root organs of the plant – just as there is with the green leaf. There is a vertical or stem-like organ (the root) and another that opens itself out towards the space around (the root cap and root hairs). Unlike in the upper plant, there is no specific structure, only a type of growth (either more vertical or more horizontal, etc.); its structure is largely determined by the surroundings. The tip of every root – like the growing point of the stem – contains meristem tissue, which means it has the capacity for continually forming new cells.

The root tips are not only extremely active, they are also very sensitive. They are sense organs of the plant. Charles Darwin described the apex of the root as being like a brain that receives impressions via the sense organs and even instigates various movements. This hypothesis is also made use of by the instigator of biodynamic agriculture, Rudolf Steiner, when he compares the plant to a human being standing on its head. In this image the root system corresponds to the human nerve-sense system and the flower to the metabolic pole with its digestion and reproduction.[6]

It is quite apparent that we must abandon the notion of a passive root that simply takes in water and nutrients mechanically. After its initial growth phase when the plant needs to absorb soluble nutrients, it can use its roots to mobilise soil nutrients in an active way. The plant is able to perceive what it needs from the soil (water and nutrients). During vegetative growth a portion of what is assimilated is given back to the soil as specific root excretions. In this way each plant species forms its own soil environment by dissolving particular soil minerals and encouraging a specific soil life. If a plant is to be cultivated in a way appropriate to its species, the question can be formulated as follows: Can the capacities of the root be influenced in similar ways to the senses and thinking capacities of human beings, by using suitable practices and nutrition? Is it possible to make the plant more 'intelligent'?

The root is an active plant organ which seeks to extend itself ever deeper and further in all directions in order to connect itself as intimately as possible with the earth. Root and earth should ideally be thought of as a unity. That would correspond far more closely to reality, especially when the mycorrhizal fungi which penetrate the roots are taken into account. These can be considered qualitatively as an emancipated root zone. The contact area between plant and soil can be increased tenfold through them. It has been shown, for example, that they can help the vine take up more phosphorous, nitrogen, potassium and zinc from the soil. In drought conditions they also help the vine obtain more water from the soil since they can penetrate the spaces that roots cannot.[7] In this way the plant can reach deeper into the soil in its active search for the nutrients it needs.

The root has a further function, namely to produce wood. Almost immediately behind the very sensitive root tip which is so open to the surroundings, the root closes itself off by creating a kind of bark and forming wood towards the inside. Furthermore, much of the internal tissue dies off in order to create transport vessels.

The creation of substance

For Goethe, everything a living being expresses mirrors its inner nature. In his *Scientific Writings* he wrote, 'Nature has neither core nor husk, she is everything at once.' According to this view, properties like form and colour express the inner nature of the plant and are not the product of some complicated biochemical factory. In order briefly to sketch out how substances are created in a plant we can draw on the alchemical principles of Salt, Mercury and Sulphur (see Chapter 12).

The plant is at its most plant-like where it creates living substance through photosynthesis in its green parts (stem and leaf), making sugar out of weighty terrestrial elements like earth (mineral salts), water and air (carbon dioxide) as well as cosmic influences like light and warmth. The substances thus formed develop in two directions:

> ❦ *Densification.* Plants form dry, earthy, solid, lasting substances (e.g. wood formation with substances like cellulose, lignin, tannins, etc.). Those substances with a tendency to solidify, mineralise and crystallise. This process of mineralisation (for alchemists this is known as the Salt process) typically occurs in the root. It can however also take place higher up, above the earth, especially with woody plants. When looked at in this way, the tree trunk and its branches can be described as 'raised up earth'. This is where the wood forming plants store their reserves for the cold season.

❧ *Refinement.* In the cosmic pole of light and warmth the substances formed in the green organs of the leaves (carbohydrates) are continually being differentiated, refined and raised into weightlessness (Goethe spoke of the 'refining of fluids'). In the process they become ever more specific to species and variety. We only need think of the wide range of rose scents, each variety having its own smell. These secondary substances (pigments, essential oils, fatty oils, etc.) are given off to the surroundings. Unlike starch, for example, they cannot be transformed or made further use of. The plant in effect dissolves itself in the flowering pole (the Sulphur process).

The green living plant (vegetative or Mercury process) thus 'dies' in two directions, downwards and upwards. In the earth by forming dense, mineralised, solid, permanent substances and a great deal of mass (wood formation), and towards the heavens by dissolving and releasing from itself strongly differentiated and highly specific substances in comparatively small amounts, while nonetheless expressing its characteristic identity in the strongest fashion (flower formation).

The huge difference between these two processes becomes clear if we compare the amount of wood coming from a hectare of woodland with the quantity of essential oil produced by a hectare of rose plantation.

Fruit formation

How then do we develop the criteria and perspectives for producing high quality fruit out of what has been described up till now? Perhaps the first thing to ask is what are we looking for in a good table fruit? It should on the one hand be large enough, juicy and fresh. It should also be a beautiful colour, have a good flavour and also smell good. A fruit should unite the vegetative tendency of the leaves (substance forming) with the qualities of flowering (flavour, colour, differentiation). If vegetative forces dominate, we will have large watery fruit that are tasteless and don't keep well. They rot easily. If on the other hand the flowering qualities dominate, fruit will often be too small but have an intense flavour and aroma. They dry out instead of rotting. Criteria based on all this can be formulated for the healthy development of a fruiting plant.

Woody plants

The vine is a woody climber. In order to understand it we need to be familiar with woody plants. With an annual plant all the substance which has been created – apart from the seed – is broken down and transformed into humus ready to support the next season's growth of new plants. With woody plants the 'stem' which grows up vertically from the earth and emancipates itself from the immediate connection to the earth, is not broken down again. It remains standing and in the autumn forms wood and bark. Only the leaves are subject to the breaking down processes that annual plants undergo. The wood forming process which with annual plants remains in the root beneath the earth, rises upwards in the case of woody plants. The foundation for new life, which the decomposition of annual plants brings about, has its corresponding parallel in wood. In both cases a living soil is created. The non-wood forming plants return most of their substances to the earth, while the woody plants 'condense them to wood'. Each woody plant can therefore be considered as a three-dimensional community of herbaceous plants growing above the earth. For this reason, the botanist Francis Hallé also spoke of a tree as being comparable to a colony of coral.[8]

In order to understand this approach even better it is necessary to compare the growth of the woody plant more closely with that of the annual plant. While in the case of the annual plant we speak of only one form of growth (arising out of the growing point) – a growth in length (also known as primary growth) – in the woody plants there is also a so-called secondary growth – a growth in thickness. That is ultimately the process which brings about the formation of wood. Added to the meristem tissue of the growing point is some further meristem tissue which surrounds the entire woody plant – the cambium. It retains the capacity to produce new cells throughout the life of the woody plant and even to close wounds (branches that have been cut off, etc.). From this we can understand how suddenly new and strong shoots can grow from the trunk of an old tree. The cambium creates wood on the inside (with xylem), and bark (phloem) on the outside.

Figure 11. The tree as raised up earth, the annual shoots like plants growing upon the tree.

The tree is therefore raised up living soil in which seeds – the buds – produce annual plants each year. This image of the tree being 'raised up earth' that Rudolf Steiner described in his agriculture course, was also used by earlier well-known biologists like Charles Darwin, Jean-Henri Fabre and Goethe.[9]

The relationship between seed and bud which has just been presented could offer an interesting perspective with regard to the regeneration of the vine – especially since regeneration via the seed is no longer possible thanks to current legal requirements in Europe. These require the European vine to be grafted on to an American rootstock.

If the buds on the tree equate to the seeds in the soil then we may ask, where are the roots of the annual plants that sprout from the tree buds? In the agriculture course Rudolf Steiner sees the cambium as splayed root, as a root replacement. There are also botanists today who like Francis Hallé, suggest that wood should be considered a community of roots of the annual shoots.[10]

While in the annual plant the flowers most commonly appear at the end of the stem, the woody plant retains a vegetative growing point with

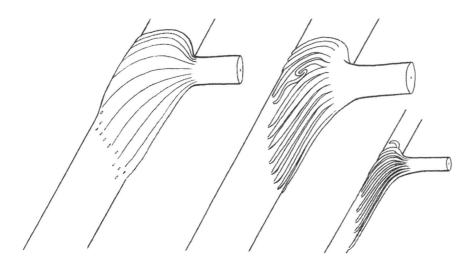

the potential of developing a further shoot. In the axle of every leaf, too, a bud forms that will sprout the following spring.

Densification and contraction (wood formation and seed or bud formation) takes place in the descending year (after midsummer). As they grow, bushes and trees create their own architectural structure. From a specific age onward, they also start to flower and produce seeds. It is thus possible with woody plants to distinguish stages of youth and maturity.

Figure 12. Three examples of plants whose twigs first form roots and then grow together as branch or trunk. The one on the left shows no clear boundary between root and wood tissue.

2

THE VINE

Jean-Michel Florin

Building on what has been discussed regarding the plant archetype, we can now try and characterise the unique quality or nature of the vine. How does this general principle of plant nature manifest itself in the specific nature of the vine? What is the unique quality or gesture of this plant? We will attempt to characterise it as phenomenologically as possible. We hope the descriptions that follow – which can surely be deepened – will serve as a stimulus to a more intimate and living understanding of the vine by making one's own observations.

A woody climber

Vine classification

The vine belongs to the order *Vitales* and the family *Vitaceae* which is morphologically close to the *Rhamnaceae* (buckthorn family). There are around 700 species in this family, most of which grow in tropical regions. They are subdivided into twelve genera of which *Ampelopsis* (peppervine) and *Vitis* are the most well known. The *Vitis* genus has only one European (*Vitis vinifera*) and several American and Asiatic species.

The wild vine and the cultivated grape vine

For botanists there is a single species *Vitis vinifera* which is subdivided into two subspecies: *Vitis vinifera* subsp. *sylvestris* (the wild forms) and *Vitis vinifera* subsp. *vinifera* (the cultivated forms).

There are some interesting differences between the two subspecies – the wild vine (*V. vinifera* subsp. *sylvestris*) grows in damp woodland, especially along stream banks and woodland edges, whereas heliophilic lianas can climb from 15 to 17 metres (from 50 to 60 ft) up trees to reach the light. They fill up all the spaces between trees on the woodland edge. Like other lianas (such as clematis) they create a 'blanket' and 'heal the wounds' of the forest edge. Their strong growth and ability to grow long shoots from every leaf node means they can very quickly fill up these gaps. They seek out forest soils that are fresh and rich in organic matter. Unlike other cultivated plants such as cereals, which in their wild state nearly grow as monocultures, the wild vine always grows where there is great biodiversity. Streamside forests are particularly known for their wealth of herbs, bushes, climbers, trees and animals. Through their red berries they have a special connection to birds.

Their natural environment – in complete contrast to where they are cultivated – is a site with humus-rich soil, shade, moisture and warmth. They live in the 'tropical' zone of our landscapes.

In the nineteenth century they were decimated by the grape louse (phylloxera) as well as fungal diseases like mildew (see Chapter 4). These

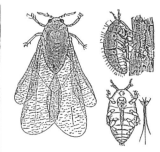

Figures 13–15. The grape louse and the damage it causes.

pests and diseases were imported from America with the American vines whose rootstocks were used for grafting. Wild vines can still be found in certain places in Europe such as the moist coastal forests of Corsica or the Basque country.

As has already been mentioned the wild vine has the interesting characteristic of being dioecious. There are male plants with pollen flowers and finely divided leaves, and female plants with seed bearing flowers and rounder shaped leaves. These formative tendencies accord with observations made with other dioecious plants (see above). The cultivated vine has lost this dioecious characteristic.

Apart from the wild subspecies we often find the so-called 'wild grapevine' growing along the boundary of vineyards. These have either grown from a stray grape seed or reproduced themselves vegetatively. Interestingly, they are often resistant to the grape louse. Could they be used to produce rootstocks? By observing such plants, we can also form an impression of how vines behave when they are not pruned.

Figure 16 (above). Leaves of the wild vine (Vitis sylvestris) *male and female. (Bournérias, et al, Le Golfe de Gascogne.)*

Figure 17 (right). The wild vine (Vitis sylvestris) *trailing over a tree in Corsica.*

American species

Take the example of the American species *Vitis riparia*, which is used to provide a rootstock for European vines. Like its European relative it also grows on the forest edge and especially along riverbanks. It clambers high into the trees, covers them with its long shoots and fills all the gaps.

The American species appears to have more vitality than the European *V. sylvestris*. Their roots don't go down so deep, however. We may therefore question the compatibility of *V. sylvestris* and an American rootstock (e.g. *V. riparia*). This question was already being posed in 1900 by Lucien Daniel, the French agronomist and grafting specialist.[1]

The grape louse plague

In the mid-nineteenth century American vines were imported into France. They had the advantage of being resistant to powdery mildew (*Erysiphe necator/Uncinula necator*) while the French vine was susceptible to this fungus. Unfortunately, the wine made from the American grapes had a very poor taste, was very bitter and sharp and was reminiscent of the rice flour used for cosmetics. This failing very quickly dissuaded wine growers from planting the American vine, at least to begin with.

Along with the import of American vine plants between 1854 and 1860, two very bad vine diseases came which up until then were unknown in Europe. One was caused by an insect that mostly attacked the roots – the grape louse – and the other by a fungus – grapevine downy mildew (*Plasmospara viticola*). The American vine is resistant, whilst the European vine is very susceptible to these diseases. Both the grape louse and the mildew spread rapidly through the vineyards.

At a congress in Beaune, France, in 1869 Laliman and Gaston Bazill recommended grafting French vines on to American rootstocks in order to prevent the destruction of vine-growing regions through the grape louse.

The method of saving the vineyards by grafting French vines on to American rootstocks was based on the inexact hypothesis that scion and stock retain all their characteristics when brought together. The near universal application of this approach, known as reconstitution,

Figures 18–21. Removal of European vines at the turn of the twentieth century.

fundamentally changed the old way of propagating vines with cuttings or runners that had been practised in Europe for thousands of years and called for the development of new techniques. This reconstitution brought many problems with it which to this day have not been solved. These problems concern the mutual adaptation of the two halves of the graft (stock and scion), adaptation to soil and climate, varying resistance to parasites, changes to the pruning regime, the change in yield and quality of the graft, etc.

When reconstitution began these issues were still unsolved. As a result, many mistakes were made and real scandals occurred. Because of poor quality it became necessary to permit the use of all kinds of additives in the treatment of the wine and fruit must.

The chemical industry was not only interested in selling pesticides against the grape louse, it also wanted to provide the additives needed to improve the wine and fruit must that came from grapes whose quality was compromised as a result of the grafting. The growing demand for sulfur and copper sulfate in the fight against downy and powdery mildew, was also in its interest.

As Lucien Daniel said later: 'By using the grafting technique to rescue our old vineyards from the grape louse, we made them vulnerable to fungal disease and pest attack.' The American rootstocks brought with them many new parasites like the grape louse and mildew and although they were resistant to them, the same could not be said of European vines.

The reconstitution of vineyards using grafted vines was also a business opportunity for cane suppliers (importers and suppliers of American vine), the rogues who mixed the wines in large quantities, as well as sugar beet producers and sugar factories in the north who supplied their leftovers to make 'sweet wine' and 'improve' average quality wine.

Vine grafting

The reconstitution of vineyards with grafted vines created many problems. Daniel investigated it systematically. It is known that while the roots of the American vine can cope with the grape louse quite well, the roots of the European vine are severely affected – hence the idea of giving the European vine an American rootstock.

The French vines were grafted on to the American *V. riparia* and *V. rupestris*, as well as other species and their hybrids. Although *V. riparia* is shallower rooting than *V. rupestris,* as with all the other American vines used as rootstocks, both have a more or less extensive network of roots that draw nourishment from the upper layers of the soil. It is different with the French vines whose roots go deeper down.

In order to thrive the American vines pump a large quantity of water from the moist and generally well fertilised soil.

Once the European vine has been grafted on to the American rootstock it receives more water and more nitrogen than it would have done through its own roots. And this despite the fact that the grafting point slows down the passage of plant sap from the rootstock to the scion.

Under normal climatic conditions the grafted European vine is thus supplied with far more water and plant sap than its own roots would have drawn from the soil.

Daniel said that by grafting the healthy European vine on to an American rootstock an imbalance occurs between the functional capacities of the different grafted parts under normal climatic conditions. The absorption power of the European vine is weaker than the absorption power of the American rootstock. Under these conditions this situation reflects a moist and fertile context for the European vine. In spring the European scion does in fact grow faster when the rootstock becomes active, than it would in an original non-grafted European vine. When it flowers, the grafted European vine produces larger inflorescences and, if the weather is fine, good fertilisation. The grapes develop quickly and are closely packed. The vegetative part too (leaves, branches, twigs) is more strongly emphasised than with non-grafted vines.

These differences of the European vine on an American rootstock as compared with the original form of the European vine can be observed under normal climatic conditions.

Grafted vines in dry years

The grafted European vine is far more sensitive than the original form to sudden changes in temperature with hot dry periods and high humidity. If the drought occurs during flowering the fertilisation of the grafted

Figure 22. Lucien Daniel (1856–1940) recognised early on the problems associated with modern viticulture.

Figure 23 (above). Vitis riparia *(New York state) growing over a tree.*

Figure 24 (right). Leaf of the American species Vitis rupestris.

vine is initially stimulated, but if the drought continues it will suffer far more than the original vine.

It is indeed the case that less **water** is then taken up by the vine. The American rootstock which is greedy for water is then unable to find enough with its shallow rooting system. If the European vine could use its own roots that go far deeper, it would have no problem finding water nor be so stressed. Added to this, the grafted point has the effect of slowing down the flow of what little liquid the American rootstock is able to obtain with its roots.

Just as the European vine has a lower uptake capacity than the American rootstock under normal weather conditions, in a drought situation it is the other way around: the grafted European vine has a greater capacity for taking up fluids than the rootstock.

Grafted vines in wet years

If, however, there is too much water in the soil, the American rootstock which can take up a great deal of water, pumps a lot of it into the vine. The rootstock suffers more from this than the scion which is always able to give off water by forming more leaves and, if necessary, new shoots. The grafted point however holds back the flow, and the water is congested in the roots. The imbalance between the American rootstock and the European scion becomes ever larger.

According to Lucien Daniel grafting results in two strong new factors of variability:

- ❦ The grafted point which filters the flow of sap and reduces the uptake of sap by the scion.
- ❦ The different functional capacities of scion and rootstock.

By growing the vine in an environment that encourages flower and fruit development – poor soil, more light, less moisture – it could be said that human intervention has changed the relationship between the wood-forming, vegetative and fruit-forming processes. During the twentieth century these relationships were changed further still by the introduction of large quantities of artificial fertilisers. These fertilisers encouraged the one-sided development of vegetative growth.

Added to this is the fact that as a result of the grape louse crisis a complete 'Americanisation' of European vineyards took place. As has been described above, these American rootstocks are not really compatible with either the European soils or the European varieties from which the scions are taken (see also Chapter 3).

Out of all this comes the urgent question: How can the wood-forming, vegetative and fruit-forming processes of this cultivated plant be brought back into a healthy balance?

The juvenile growth stage

A vine seedling of *Vitis vinifera* begins its development by sending down a tap root; afterwards an upright and vertical stem with no tendrils and two seed leaves is produced. The root can go down very deep into the earth; well-grown roots have been found to go down 15 m (50 ft). The American variety used as a rootstock today does not go down as deep. They are also the product of vegetative propagation and therefore produce only adventitious roots, that is, roots that form from non-root tissue.

The seedling of the European vine then develops a spiralling series of eight to ten leaves (repeating at $^2/_5$ intervals around the stem). This is a typical gesture of vegetative growth that reflects the sun's influence. Buds are formed in the leaf axles with two protective leaf scales.

In the years that follow the vine plant undergoes further changes:

- The spiral arrangement of leaves changes to an alternating arrangement, one that is comparatively rare in the plant world and which comes over as very regimented. Dieter Bauer showed that this arrangement is formed by a second force (gravity).[3]
- The vine develops tendrils, loses its verticality and uses its tendrils to find something to hang on to. It becomes a true climber.
- Tendrils and flowers develop in the leaf axles in an

alternating pattern with the leaves. The flower shoots and tendrils at first grow skywards and then conceal themselves beneath the leaves.

🍇 The tendrils appearing with the flower stems seek ways of fixing themselves.

🍇 The vine flowers for the first time when about six years old. It then reaches its mature stage.

The mature grape vine

All the buds on the vine plant proceed to grow. This is not the case with trees. The inhibition of growth can be seen most clearly in conifers where the leading bud forms the longest shoot and prevents the side shoots from developing. In the vine this inhibiting tendency is greatly reduced for every bud can produce a long shoot. That is why the vine can grow in all directions and rapidly fill a large space.

Like all climbing plants the vine is very mobile and opens itself to its surroundings by renouncing a characteristic form of its own – which most other trees possess. It is precisely this openness to its surroundings which allows the vine to be cultivated in so many different ways (espalier, pergola, goblet, etc.).

Figures 25 and 26. The spiral growth pattern of the young vine grown from a seed. This stage is missing with vegetative propagation.

In times gone by vines grew to a great age. Reference was often made of vines more than a hundred years old. Some of the old vine stocks reached a circumference of 50 to 60 cm (20–24 in). In 1885 a vine stock was recorded in Portugal that had a girth of 2 m (7 ft) and whose vines covered an area of nearly 500 sq m (5000 sq ft). In 1864 this vine stock produced 745 litres (195 gallons) of wine.[4] Modern vines by contrast rarely live more than 30 to 40 years. This is a further sign of how weak today's vines have become as compared with their earlier potential.

A large number of shoots will develop from the previous year's buds of a vine stock if left unpruned.

The structure of a first-year vine shoot is very unusual and is made up of two parts. The first part consists of a 'preformed' section that was formed in the previous year's bud. This generally has about eight leaves and two or three flower stems and primarily draws on plant reserves in order to develop. Then, after the really short internode of the preformed section, a new section which was not preformed develops. This means that the shoot grows beyond what was predetermined the year before. New leaves are formed which reflect the altered environmental conditions. This new section forms some 20 to 25 leaves and, instead of flowers, grows tendrils.

It is worth noting the unusual nature of this growth: in most plants the vegetative and flowering processes are polar opposites. Either the flowers develop at the end of the stem and the vegetative processes are inhibited or devitalised, or, as is the case with many trees, two kinds of shoots are formed: a longer woody shoot representing the vegetative process, and a

Figure 27. The modest flower of the vine in spring.

shorter flowering shoot representing the flowering process with its growth-inhibiting effect.[5] Fruit trees, such as cherry and apple, demonstrate this particularly well.

The vegetative growth of the vine is so strong that it cannot be held back by the flowering shoots. In connection with flowering this would correspond in the case of an annual plant to continued growth – from out of the flower a new stem would develop.

It is also unusual in that there are two dormant buds in every leaf axle instead of just the usual one, which can open the following year. One of the buds produces a new long shoot and the second, usually smaller bud, produces a short, so-called side shoot that produces some leaves and occasionally small flowers as well. This once again demonstrates the uninhibited vegetative power that enables the vine to grow out in all directions.

There is another special quality belonging to the fully-grown shoot, namely the relative positions of flowers and tendrils. The flowers, which interestingly enough appear together with the tendrils, grow in a special sequence. Starting from the base there is a succession of three leaves at 180°. Then comes a leaf with a flower opposite. This is followed once again by two further leaves but now the second leaf comes with a flower opposite.

Once two or three flower stems appear the rhythm continues, though this time with tendrils instead of flowers. Leaf and tendril shoot alternate followed by two leaves and a tendril shoot. Once again we can see that although the flower stems are pushed to one side (and play a somewhat

Figure 28. The flower head and tendril form the end point of one unit.

Figure 29: Vine shoots are composed of several basic units. In the sketch we see the alternation between white and grey.

subordinate role) there is no devitalising or inhibiting of the vegetative process. It is amazing that the flowers of the vine and ultimately its fruit are borne beneath the leaves and shoots.

Botanists have had many discussions about how to explain this special growth habit. According to one hypothesis it is a form of sympodial growth. This would indicate that a one-year shoot is made up from a series of units or shoots. The first unit produces three to five leaves followed by a flower stem which concludes the growth. The next shoot forms a leaf which is also followed by a flower stem and the third shoot forms two leaves which are again followed by a flower stem and so on until the flower stems are replaced by tendrils. Each unit comes to an end with a flower stem or a tendril.

If we try to follow this growth dynamic we discover once again the special vegetative power of the grape vine – each shoot is succeeded by an equally strong shoot, and so on seemingly for ever.[6]

The leaf

Vines, trees and almost all woody plants do not have a complete leaf metamorphosis. They only show the first stages. The first leaves are mostly small and undifferentiated and the next leaves are much the same. There is no contracting phase. The form of the leaves remains largely under the influence of the first two formative forces – expansion and filling out, with a more or less strong differentiation depending on the species. This again shows how the vegetative process is dominant. The last stage of coming to a point is only found on the tiny leaves on the flower stems.

The vine leaf grows out on a long stem. The leaf blade tends to be round and not very differentiated. A diversity of leaf forms has evolved in the vine as a result of human plant-breeding activity. These range from the virtually round (in the case of the Gewürztraminer) to the very finely divided leaves of, for example, Cabernet Sauvignon. The form of leaf and entire morphology of the various varieties reflect their different qualities. It is possible by observing these forms to find varieties that suit the particular site and climatic conditions. There are varieties for example with finely divided leaves that are better suited to warm and sunny situations. Close observation will also be a great help for choosing the most appropriate variety and rootstock.

To summarise we can say that the leaf form of the vine is rather typical of a plant that has great vegetative strength. The American varieties used as rootstocks tend to have less structured leaves. Sometimes their leaves are asymmetric or have a bulging form – both of which signal an unstructured vegetative power. Their grapes also don't produce a good wine which is why European vines are grafted on to American rootstocks.

A closer look at the leaf arrangement on the vine and that of a comparable maple leaf reveals an unusual gesture. The leaves appear to have a slight spiralling tendency (torsion), as if they had twisted themselves.

The leaves start off a light pale green colour, which deepens to a strong dark green in order then in autumn (like many deciduous trees) to take on rich yellow, orange and red tones. This is where once again we find a close affinity between vines and trees. The colours described often appear when vitality declines, revealing qualities which otherwise occur with flowering and fruiting. The leaves appear to ripen.

The leaves of flowering herbs rarely take on colour, they remain more closely connected to terrestrial forces. Only when they have gone through the process of metamorphosis, freed themselves from terrestrial forces and opened themselves up to the light and warmth of the cosmos, do they flower and take on such yellow, orange and red coloration. Each grape variety has a different way of ripening and colouring, and depending on its situation and cultivation, a different range of colours. (There are also varieties which take on an intense red colour like the Meunier variety.) All these things are worth observing in detail especially with regard to the qualities that can be deduced.

Tendrils

The vine tendrils as well as the flower cluster (the two organs belong together) are today considered to be transformed stem. This is also confirmed through the fact that they often bear upper leaves. How should such a modest organ as the tendril be understood? Vine tendrils are sensitive, vegetative plant organs which initially grow towards the sky, weightlessly defying gravity. But they are soon on the lookout for something solid and earthly (that is, the opposite of weightlessness) to hold on to. The tendril is a stem permeated by the spiralling principle of the leaf. It is also very sensitive – like an airborne root that seeks out solid objects. This is where the climber-characteristic of the vine shows itself. The tendril first starts growing upwards but very soon begins to explore its surroundings. So long as it has not found something to cling to, it remains youthful and extremely sensitive, and has a great deal of mobility and plasticity. In its young state it can even move around before eventually rolling itself around something and becoming fixed and woody. Tendrils that don't find anything to cling on to die and soon fall off.

The exploratory gesture of the tendril has more in common with an animal than a plant, which, although open to its surroundings, very rarely sets out to explore. The vine tendril branches into four parts in a binary fashion. This is a very primitive form of division and typical of algae.

Flowers

Instead of forming coloured and 'devitalised' plant organs (petals) that open cup-like towards the sky, the vine produces a flower cluster made up of many small green flowers that remain hidden beneath an overhanging leaf. The flowers seem so small that the flower cluster might be thought of at first sight as being perpetually in bud. The very modest individual flowers are only about 5 mm (¹⁄₅ in) across. They are protected by a small green 'hat' that opens up at the base and eventually falls off. The flower then stands with its stamens and ovaries, naked and alone. It retains a strongly vegetative quality and gives out a powerful aromatic scent.

The flower cluster grows together with the tendril. We often find tendrils with the beginnings of a flower cluster, and there are also many in-between stages. The association of tendril and flower is very rarely found among plants; most tendrils are metamorphosed leaves or stems. What does this mean for the vine?

Tendrils and flower clusters remain at a vegetative phase and their growth does not end at the top. Unlike other well-structured and organised flower clusters with their characteristic and colourful blossoms, the flower cluster of the vine – referred to by botanists as thyrse – has far more vegetative characteristics. The flower clusters appear beneath the green leaves in the lower section of the shoot. They are made up of a large number of tiny flowers without a strict organisational structure. The cluster has numerous flowers which branch out on four levels, which means an intense ramification. This form of branching appears to be far less organised than in *Umbelliferae* or *Compositae*, for example, whose

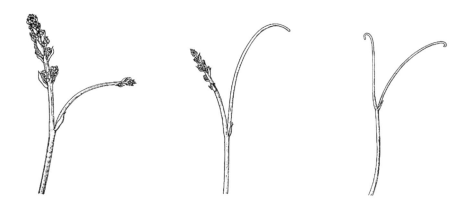

Figure 30. In-between stages showing the homologous relationship of tendril and flower.

single flowers reflect the form of the entire cluster. Here the flowering or astral forces are much stronger.

To sum up, we can see how flower clusters and individual flowers are strongly determined by vegetative processes. At the same time, the fine scent of the vine flowers reveals a true flower quality even though in terms of form and colour it is very restrained. It suggests that the flowering potential (colour, form, structure, but not scent) is held in reserve. After it has flowered tiny green fruits develop.

The ripening vine

Figure 31. The vine flowers express a primitive vegetative structure, dichotomy, colourlessness and minimal structuring.

With the start of the ripening process the vine gradually reduces its expansive and vigorous growth and focuses on an internalised process. The fruits which to begin with are green, start to ripen. Secondary plant substances begin to form and refine themselves (pigments, aromas, etc.).

New processes, especially those connected with warmth and light, now start to have an influence. The fruits with their four pips slowly begin to ripen, the leaves start to colour and the annual shoots turn woody. Leaf colour ranges from yellow through to deep red depending on the way it is cultivated. Because of this it is easy to distinguish from afar the different plots in a vineyard. Seasonal variations also affect the colour. A warm moist autumn can hold back the colour, too much nitrogenous fertiliser can have a similar effect.

The colour changes of the vine can be compared to the autumn colours of trees. In both cases we have plants whose flowers are very modest and yet through their coloured leaves in autumn express a kind of 'flowering'. Colours come about as plant vitality declines, which with many woody plants occurs only in autumn once the light and warmth of summer has come to full expression.

It could be said that the vine only flowers in the autumn when the fruit and leaves take on colour. It is also the time of year when the vine becomes most noticeable in the landscape. The colour of the vines has an effect on our soul as strong as that of a meadow full of flowers, albeit of a different nature.

This tendency reaches a climax with the red grape varieties (*Vitis viniflora* var. *tinctoria*). These varieties in particular, which have black grapes and red flesh (all other varieties have white flesh), possess special healing qualities for the circulation of blood. These are varieties whose leaves take on an intense red colour in the autumn. They are also known as 'dye grapes'. Examples include the many varieties of Teinturer grapes, grown widely in France, North America, New Zealand and Australia, and the Fumin d'Italie and Colorino de Toscane from Italy.

Grapes

Even in its fruit the vine retains its vegetative character. The fruits are very juicy, contain a lot of water and remain fresh for only a few days. They have the tendency to rot already on the vine.

The grape is a 'cold' fruit. Eating grapes in warm weather is experienced as being very enjoyable, when it is cold however it is often difficult. One has the impression that grapes strengthen coolness. This contrasts with say, the apple, which is more of a 'warm' fruit.

Figure 32. Grapes ripening on the vine.

The fruits are to be found where the flower clusters were, beneath the upper leaves. This contrasts once again with most other fruits which appear above the leaves at the end of a shoot where vitality has been withdrawn. Grapes are thus produced in a plant zone that is full of vitality, which is still strongly permeated by the stream of life rising up from the earth. This vegetative quality has a spring-like characteristic. It is one more often associated with early fruits like strawberries and cherries, which although very juicy are very hard to store. In the autumn there are usually harder and dryer fruits that have undergone the devitalising process, take longer to ripen, and keep well, such as apples and quinces.

The modern breeding of vine varieties which promotes pruning and the American rootstock has increased the watery tendency of grapes. By contrast, an unpruned vine forms many bunches of small grapes which hardly move and allow a lot of light and warmth to come among the grapes (see Chapter 6).

It is perhaps due to this vegetative characteristic of the grapes that we can understand how it is only through the process of fermentation, itself a form of ripening aided by bacteria and yeasts, that the many flavours and substances manifest themselves.

There are also other traditional forms of post-harvest ripening such as the drying of raisins. On the one hand grapes can be ripened in the cellar, in the shade (as has been mentioned, with the help of terrestrial organisms), and on the other by drying in the warmth and light of the sun: two polar opposite processes.

Until the grapes begin to ripen the fruit remains vegetative – it can still continue assimilating through photosynthesis, grows strongly, contains many acids and forms very little sugar. As soon as it begins to ripen, however, the content of sugar substances (glucose, fructose) rapidly increases. Within a mere month the content rises from 1 per cent to 20 per cent by weight of fruit.[7]

At the same time as this transformation into sugar substances, the shoots become woody. The influence of light and warmth brings about a densification or ripening process. The grapes take on colour; this means they lose some vitality and form secondary substances like pigments, sugars and tannins.

It is only now that a process of differentiation begins (in many plants this already occurs with flowering) and wood begins to form. In the centre of the grapes, seeds are formed from which oil is produced. Light and warmth encourage this process of contraction.

Another very interesting aspect concerns the proximity between grapes and the yeasts that enable winemaking. It was thought for a long time that such yeasts were already present on the grapes. It seems however that in the course of time vintners bred the appropriate yeasts.[8]

Some fruits ferment very easily and the grape is a good example of this. But there are also fruits that virtually never ferment, such as blackcurrants (*Ribes nigrum*) or bilberries and cranberries (*Vaccinium myrtillus, Vaccinium vitis-idaer*), which contain substances that actively prevent fermentation. In this regard the entire bilberry plant family, the *Ericaceae*, stand as polar opposites of the vine. While the *Vitaceae* are largely tropical plants that need warmth, the *Ericaceae* are moorland plants thriving far up in higher, northern latitudes. They are also morphologically and physiologically opposite to one another. While climbers grow outwards producing a lot of mass, have a tendency to accumulate water and always remain in a youthful state, the *Ericaceae* form small, compact, ancient-looking 'little trees' that are very woody and bear small, leathery leaves; their whole growth appears stunted.

Wood formation

It is worth closely observing how the wood of the vine ripens during the time when the grapes ripen and colour (from August onwards in the northern hemisphere). Those buds are being formed that will produce new shoots in the next year. Three processes are at work here: ripening, densification and internalisation. The plant is now no longer engaged in growing (outwardly) but is drawing itself together (inwardly). Substances are being transformed, refined or dried out. The tendrils have become hard, the old bark has dried out and begun to crack and exfoliate. The drying out remains superficial since in the vine the process of wood formation does not go as far to the centre, as it does with trees. There remains a green cambium layer and a soft core which makes the vine particularly responsive to pruning (see Chapter 15).

The vine's gesture

In its essence the vine is a plant connected deeply with the earth, which, through its process of forming wood, lifts the 'earth' skywards. In doing so it can become very old – over a thousand years. As a pronounced vegetative plant, it is open and receptive towards its environment as is also shown by the fact that it develops tendrils. The weak and vegetative flowering process is left to one side and the vegetative power of the vine invested directly in the process of fruit formation. Wine growers use all means at their disposal to guide this vegetative process, such as pruning for example. Vine shoots and even the tendrils as transformed stems, are seen in the tension between two polar forces, between the vertical up-striving and the force of gravity which bring the shoots into the horizontal. The vine always wants to grow upwards without being able to really manage it. It needs the support of trees upon which to climb – an image of continual longing to ascend and reach the light.

In comparison to many other plants, there is no devitalising contraction expressed in the leaf of the vine. On the contrary, the vegetative power is so strong that instead of remaining dormant, even the secondary buds in the leaf axles of the current year's shoot produce side shoots. In some cases, further shoots develop from these side shoots. In late spring the grape vine gives the impression of being a plant that wants to grow out in all directions and fill all the available space without being able to stop growing. The vine bristles with vegetative energy and it continues to channel this energy right into the fruits, which retain their youthful, watery, terrestrial characteristics until they begin to ripen. Rudolf Steiner described this process in the following way:

> What other plants reserve solely for the young germinating seedling, all the growth power which is otherwise saved for the young seedling and is not poured into the rest of the plant, is in a certain way poured into the flesh of the grape by the grapevine.[9]

This makes it possible to understand how the grapevine, more than any other kind of fruit, reflects its environment over the course of the year. The flowering process in the vine as a process of differentiation appears to be strongly displaced. It takes hold when, as the fruit begins to ripen, many special secondary substances develop in the skin and seeds of the fruit (polyphenols and the acclaimed resveratrol). These substances are closely related to tannins and reflect how the flowering process, which now becomes particularly intense, acts to restrain and transform the vegetative power. At this stage the leaves also respond to the flowering impulse by developing strong colours.

The transformation of the grapes continues through the process of fermentation which the vine (or rather the grape) enters almost of its own accord. A whole range of yeasts and bacteria then help to produce the most varied array of aromas, colours and flavours. Seen in this light, the vine not only 'flowers' in the autumn colours of its leaves but also in the ripening of the grapes and their alcoholic transformation into wine. One speaks of wine having a specific bouquet which means 'a bunch of flowers'.

The formation of wood, which also takes place in autumn, is the second aspect of how substances are ripened. It enables the vine to control its one-sided vegetative power so that with renewed vigour it can climb higher the following year.

Sketching the gesture of the vine in this way demonstrates how the plant develops a strong vegetative power which is not curtailed or transformed by the flowering impulse – at least not till after the actual moment of flowering. This focus on vegetative growth, however, makes it especially sensitive and open to external influences. This helps us to understand why, for instance, the vine is particularly susceptible to fungal diseases. Grafting it on to the American rootstock, which develops an even stronger vegetative power, increased this susceptibility further (see 'The grape louse plague' p. 40).

Rudolf Steiner made an interesting comment regarding the effect of the vine on human beings in a landscape:

> The atmosphere created by the vine also helps to balance out bad effects. As you know, lime blossom is quite powerful, and walnut trees also have a powerful scent; this has more of a vitalising effect on the [soul]. And the atmosphere created by the vine has more of a vitalising effect on the I. So there you get a powerful effect also on the higher [parts of the human being].[10]

This quotation draws our attention towards another special aspect of the vine. It is a plant that continually longs to climb higher, towards the sky, to the realm of warmth and light. It can however never reach this goal. It has to rely on its tendrils year after year and try to climb up again and again. To experience this unusual growth habit more fully, we may compare it to a cereal such as rye which, thanks to its fine stem, is able to grow straight up; or perhaps to a conifer which grows in the vertical direction from the very beginning.

This extensive exploration of the vine's nature can provide us with many perspectives useful for renewing the cultivation and breeding of the vine. A few of these are briefly listed in what follows; some will then be entered into more deeply in the later chapters of this book.

Cultivation and habitat

The site

Originating in a very diverse environment, the vine was removed from its natural situation and ultimately cultivated as a monoculture. A diverse green soil covering and the avoidance of deep cultivation are the first steps being taken to improve the growing situation of the vine (see Chapters 5–7).

It is also worth getting to know each growing plot better in order to find individual varieties that are more suited to the site and to particular cultivation techniques.

All these things will remain single, unrelated measures, however, unless the vineyard in question is considered as an agricultural organism. Without an almost closed cycle of substances and forces a healthy agriculture will not be possible.[11] Animals are also important for a healthy farm. Apart from this the wild vine is a plant with a very intimate connection to the animal world and especially to the birds that eat and distribute its seeds.

How can they then be (re)integrated into the vineyard? By setting up islands of biodiversity or including animals in the vineyard there has been some success in this direction (see Chapter 6 and Chapter 8).

The soil

How can the vine be provided with a living, organic soil, so that with the help of soil mycorrhiza it can actively seek out the nutrients it needs – and without forcing the vine into excessive vegetative growth? Biodynamic fertilisation can be very helpful in bringing the many dead vineyard soils back to life (see Chapter 9).

The dead wood can also be a help towards achieving this goal. In order for it to rot down, many insects are needed, such as ants whose formic acid plays an important role in transforming dead wood into good humus. We should also think of the bees and their vital function as pollinators. How can such insects be encouraged?

Cultivation

Through cultivation of the vine, a vigorous and high-climbing liana that produced new wood each year was bred into a little tree (almost like a bonsai). How far should this 'reduction' of the vine be taken and what consequences are there for its health? By shortening the stem so drastically, the grapes which would otherwise have grown high up in the air, are produced close to the ground and therefore in completely different surroundings. Down below, vegetative forces dominate and there is less influence from the cosmic forces of warmth and light.

The correct relationship between wood, leaf surface and fruit

In earlier times vines were planted closer together, but as a result of wider spacing through the introduction of machines, the ratio of wood to leaf surface was significantly reduced. However, it is the old stocks that increase sugar and flavour.

Pruning

In order to prune the vine in a way that respects its nature, we have to understand how the sap flows in the vine stock. There is also the fact that vines sprout again very readily but that pruning cuts heal poorly (compared to trees). The person pruning the vine must understand this (see Chapter 15).

It is particularly important not to prune the vine while it is growing, especially the shoot tips. These should be retained wherever possible by, for example, tying them in (see Chapter 18).

Encouraging upright growth

The vine is continually trying to find an upright stance in order to overcome gravity. How can we support this gesture? The application of biodynamic horn silica can help. Are there other possibilities too?

Encouraging the ripening process

To cultivate the balance between growth and differentiation (ripening), there are various biodynamic measures to encourage the ripening of the grapes as well as the leaves and wood, such as horn silica and tree paste.[12]

Growing from seed

The vine has been propagated using vegetative methods for a long time. It is known that vegetative propagation eventually exhausts the plants and causes increasing numbers of pathogenic mutations. The plant can only achieve true regeneration via the seed, that is by going through a process of concentration.

Monks used to be the ones who undertook this regeneration. Grapevine seeds were sown out and the most interesting ones chosen for subsequent vegetative propagation. By this means they regularly created new diversity instead of continually reducing it, as is done today.[13]

Added to this is the fact that of the more than five thousand known varieties of grapevine, less than a hundred varieties are grown in the world. In 2010 half of all the vine plantations were planted with only 15 varieties.[14]

Orchardists who propagate their trees solely by grafting experience similar problems. Trials have been undertaken to regenerate varieties via the seed. There would also be the possibility of pursuing the bud route – which represent a similar stage for perennial plants as the seed is for annuals. The first steps have been taken in this direction in relation to the grapevine.[15]

Missing juvenile growth period

Another aspect of vegetative propagation is that the vine always begins growth in its 'adult stage'. The vine plants that are sold today are intended to be productive as soon as possible. What are the consequences of missing out on the healthy juvenile stage in which the vine enters into an active relationship with the local terrestrial and cosmic environment? The vine plant produced via vegetative propagation does not form a tap root but only adventitious roots. To a certain extent the 'earth-sky' axis is missing. It also doesn't form a vertical stem around which a spiral of leaves develops.

Grafting rootstocks

Grafting on to American rootstocks throws up many questions. Firstly, all our current rootstocks in Europe come from America, a continent with completely different local conditions whose specific nature must be considered. America is a continent with special and very strong terrestrial forces, which create unique growing conditions.

The American rootstocks tend to encourage vegetative growth. American rootstocks and European scions contradict one another in relation to their characteristic properties. Every plant has a subtle relationship with its surroundings and retains this as a 'memory'. This has been demonstrated by recent epigenetic research and is also a property Rudolf Steiner referred to in 1924.[16]

There is also a very limited range of rootstocks to choose from and this likewise greatly reduces diversity.

A paradigm change: from controlling to accompanying the vine

Rudolf Steiner's indications were never intended as dogma but rather as an encouragement to observe more broadly and more deeply. His suggestions enable us to ask about the health and future development of the vine in a more fundamental way.

The Goethean approach helps to develop a personal relationship to the plant. The plant is not simply material, but a living being that one can meet and enter into conversation with. Such a relationship is also necessary in order to accompany and care for the vine during the course of the year with appropriate biodynamic treatments. The wine grower can in this way try to perceive the particular mood and 'personality' of each plot of ground in their vineyard, with regard to site conditions, vine variety and history of the site. Such intimate knowledge of vineyard plots also makes it possible to work in a more pre-emptive way. The objective is therefore to accompany, support and strengthen the vine and not to control it. Such a relationship between human being and plant will ultimately mean that the vine is not simply trained but raised or elevated, in the best sense of the word.

PART 2
A New Impulse

3

PATHOGENESIS: THE GRAPE LOUSE PLAGUE

Georg Meissner

At this point it is worth taking a look at medicine. How is an illness or the context of an illness viewed? To the long-established concept of 'pathogenesis' (origin of illness) a second has recently been added – 'salutogenesis' (origin of health). In the words of Michaela Glöckler, 'a new, health-oriented approach to medical research was developed, called salutogenesis ... [which is] understanding how to maintain health and prevent illness in connection with medicine and education.'[1]

In this chapter we will consider the conditions that allowed the grape louse plague to spread so rapidly among European vines in the nineteenth century. In the next chapter we will look at how the biodynamic method can restore the vines to health.

In the mid-nineteenth century the inadvertent introduction of the grape louse from America nearly wiped out all the grapevine plantations of Europe. As a solution to this problem the practice of grafting European vines on to American rootstocks was introduced. Since then virtually all grape vines grown across the world have been grafted. Like today, numerous fungal diseases were also brought in from other continents.

But is the European vine constitutionally incapable of resisting the depredations of grape louse? What about the European wild vine which continues to grow, for example, along the Rhine's wooded riversides and is resistant to the fungal diseases we are familiar with?[2]

Apart from the obvious and urgent need to research questions – such as whether there are European vines that are resistant to grape louse, which species are resistant to which fungal diseases – it is also necessary to reflect on the question, why did the European grape vine become so susceptible? Had it perhaps become so weakened that it was unable to resist the new pathogen? Perhaps it was not the European vine itself but the weakened state it was in that made it so susceptible to the American grape louse?

This then raises further questions. Was the grape louse already endemic in Europe prior to 1860? What caused it to become so virulent and 'successful' after 1860? There are a number of reasons for thinking that, already by the mid-nineteenth century, the European vine must have been in a weakened state.

Vegetative reproduction

It is known that by going through the seed state a plant can rejuvenate its forces. Since the inherited characteristics remain unchanged with vegetative propagation, the plant has no possibility of adapting to an evolving environment. It is only through generative reproduction (sexual pollination) that a new combination of inherited characteristics can occur. The European vine has been propagated vegetatively for hundreds of years because it readily mutates, making it difficult to retain carefully bred varieties. This one-sided vegetative propagation means that a process of regeneration can only occur in a limited way via the cambium layer. A genuine renewal or regeneration of the vine can in fact no longer take place today through breeding. Nowadays there is only maintenance breeding and its objective is to retain the qualities of a particular clone or vine variety.

Should we then not pluck up courage and bring the vine once more into a generative cycle? It is after all done in the breeding of fungus-resistant vines. These become resistant not only as a result of crossing European and American varieties, but also because these vines have been allowed to go through a generative process.

Mono-cropping on poor soils

Vineyards were often planted on poor soils and in places where it was difficult to grow other crops – frequently on steep slopes with very little soil depth. Such one-sided vine cultivation has been carried out in Europe on poor and shallow soils, often for hundreds of years, without any true crop rotation. The question arises as to whether, by the end of the nineteenth century, the soils were exhausted.

Changes in cultivation techniques during the nineteenth century

Reading in the agricultural archives and books from the eighteenth and nineteenth centuries, we find a new approach based on the theory of fertilisation being introduced. Guano from Chile was imported as nitrogen fertiliser. Many vineyards which had previously been managed by hand were now being mechanised. Horses and oxen were increasingly used as working animals in the vineyard.

To formulate the question posed once again more clearly: did the grape louse only arrive around the middle of the nineteenth century, or did the changes in cultivation practice enable an already present insect to find fertile ground to multiply?

The response to the grape louse

What was the response to the grape louse? The approach was primarily pathogenic. Efforts were made to get rid of this pest, but these led to a further weakening of the vine. Vine cultivation provides a very early example of how the arrival of a pest can result in a great deal of genetic diversity being lost. Until the grape louse arrived every wine grower was also a breeder. They used their breeder's eye to develop and propagate their own vines. This meant that plants were being developed on site over many years and even centuries by the people living there, and were accordingly well adapted.

As a result of the grape louse plague, however, breeding was increasingly given over and ultimately transferred in its entirety to plant breeding institutions which were frequently under a state contract. Wine growers lost their 'breeder' connection to their crop. The institutes selected vines systematically and propagated them as clones on a massive scale. Where previously thousands of different vine varieties were developed out of the intuition of individual growers, the road ahead was led ever more strongly towards monoculture. Thus, for example, in the case of the Riesling vine there are now only about twenty clones available (there are many others still, but no more than twenty that are widely available around the world). And there are about five readily available clones of the American vine used as rootstock. There are of course others too, but only in small numbers. Added to this we often find that the vineyard is replanted with the same clone on exactly the same rootstock. There is frequently only one variety of clone on one and the same rootstock throughout a vineyard – an entire vineyard has in effect only one plant. And the clone has been bred to be high yielding.

As has been mentioned, the key measure taken against the grape louse was to graft the European wine varieties on to an American rootstock. The leaves of the European and the roots of the American vine have shown themselves to be resistant to the grape louse. The creation of a new plant conglomerate out of American and European plants has until now 'saved' the wine industry and virtually all the world's vines are now grafted.

But could it be that this approach was too one-sided? Did some vineyards show resistance? Was grafting on to American rootstocks the only solution? Were there, or indeed are there, alternative ways of intervening that do not have such a drastic effect on the vine's nature?

Is there any plant material available today that has not been grafted and bred for high yields? Are there perhaps old vine plantations somewhere in one of Europe's wine growing regions that have not been grafted but are still maintained as local varieties?

Some ancient non-grafted plants can still be found growing high up in a remote corner of the Mosel region in Germany. Implementing the regulation in those far off places by rooting out the European vines and replacing them with grafted vines would have been too arduous in those days. Projects are now being instigated in official circles to propagate these varieties. There is also the question about whether these plants will suit today's climatic and technical requirements. However this may be, Rudolf Steiner saw grafting merely as a short-term solution.

There are many other questions to explore here which could be the subject of a new research project.

4

SALUTOGENESIS: RESTORING THE VINE TO HEALTH

Georg Meissner

In recent years interest in biodynamic wine cultivation has grown dramatically. Reports and direct practical experience of the positive influence of biodynamic practices on soil structure, plant health and above all the quality of the wine, has meant that many wine growing operations now apply the biodynamic approach or are converting to it. Businesses that have converted directly to biodynamics from their previous conventional approach include increasing numbers of highly respected and famous French vineyards in Burgundy, Alsace, Champagne, the Rhone valley, Bordeaux and other regions. This trend can also be observed in other European countries, such as Germany, Austria, Italy, Spain, Portugal and Switzerland, as well as in North and South America, South Africa, New Zealand and Australia.

The following reasons for converting to biodynamic viticulture are often given:

🍇 Many winegrowers have found that biodynamic agriculture offers space for human engagement. By expanding the field of observation from soil, plant and animal life through to the cosmos, and by integrating it within a broader totality, for example the individuality of a farm, the work immediately takes on a more creative quality. These positive experiences often lead to an interest in the anthroposophical background.

❦ Other wine growers for whom a continuous improvement in wine quality is the prime motivation, are convinced as soon as they taste biodynamic wines. Again and again we hear descriptions of how it brings out the rich subtlety and the local *terroir*, or typical quality, of these wines.

❦ There are also many wine producers who increasingly renounce the use of oenological aids in their wine production, such as pure yeasts, enzymes and enhancers. They perceive the use of such aids as contradicting the principle of the biodynamic organism and individuality. Their goal is to produce wine that expresses the full character of the *terroir* and is subject to as little 'cellar' influence as possible.

Yet others look to an improvement in good agricultural practice through the application of biodynamic preparations:

❦ Improved soil structure, intense root penetration and a high level of microbial activity.

❦ Improved growth vigour on the site.

❦ A more balanced pattern of growth made visible by increased open foliage and looser grape clusters.

❦ Reduced use of pesticides (including copper) as a result of applying biodynamic preparations, along with various teas and plant extracts, is decisive for the future of their soils.

Finally, there is also a group who hope to improve revenue by converting to biodynamic cultivation.

With the *terroir* as an integral component, viticulture observes a central principle of biodynamic agriculture: the principle of agricultural individuality. It is something which in other fields of agriculture has first to be introduced. This close predisposition of viticulture and biodynamic principles means that it is able to respond sooner and more effectively to its practice.

Biodynamic impulses for the future of viticulture

In 1924 Rudolf Steiner gave his agriculture course in Silesia, Germany. Farmers had often begged him to give such a course. They were concerned about soil impoverishment, the increasing prevalence of plant and animal diseases and the declining quality of food. At that time the grape louse threatened the entire European grape crop. In his course, Rudolf Steiner referred to wine growing as an example of this decline. He spoke about the 'loss of traditional and instinctive knowledge'. The connection between terrestrial life and the forces of the planets was no longer considered. This lack of insight had consequences.

During one of his lectures to the workmen at the Goetheanum, Rudolf Steiner was asked whether an American clover variety, which flowered throughout the year, could be planted as bee fodder. He replied:

> The bees get all worked up and are able to work harder for a time.
> But one thing you should watch very carefully. Don't introduce
> something completely foreign to the bees, since bees, according
> to their inner nature, have become accustomed to and are tied to
> a specific area or region [...] you won't be able to impart any long-
> term benefit by giving nectar derived from regions foreign to them
> ... [The bees] will need to expend extra energy in converting the
> nectar within their bodies, where they will experience turmoil as
> they attempt to convert the nectar into a form like the form it takes
> in the place the clover came from. For a few years, you might be
> successful, but there will be real trouble later on.

He then refers to the grape vine. Steiner briefly describes the grape louse problem, how helpless people were towards it and how with the American-European plant construct that was created, viticulture became Americanised:

They were forced to give up the European vines entirely and Americanise the entire process of growing grapes. Thus viticulture became entirely transformed and evolved into something very different from what it once was. **In many** regions it has become something very distinct in many ways.[1]

What does it mean, however, for the European vine to have its 'true' roots ripped out and replaced by American ones? What effect does this have on the human cultural impulse? The European vine has after all been cultivated by human beings for thousands of years. The vine is pruned each year. Before starting to prune it is good professional practice for wine growers to stand in front of each plant and observe it quite precisely before making a cut. They must ask themselves how the sap is flowing, which direction the plant needs to grow in, how vigorous it is, and many other aspects. There is an intense and penetrating process of observation (see Chapter 15). Out of a climber, which naturally clambers up forest trees, a cultivated plant has been created. It is only through this human intervention that the plant is able to bear so much fruit. This kind of intervention did not take place with the American vine.

We can understand what this means by looking more closely at the vine. The American vine is normally stronger in its vegetative growth and has a larger amount of foliage. Its orientation is towards the development of mass. The European vine, perhaps as a result of a thousand years of cultivation, is more penetrated by form: this in turn has an effect on wine quality.

What then does grafting mean today for the cultivation of the grape vine? Does the European vine want to become an American vine in the way bees do when they feed on clover from America? Are the more powerful vegetative forces transferred to the European vine from the American rootstock?

Here we need to look at today's propagating practices. We need to consider how a young vine comes into being – and how wine growers, including biodynamic ones, use them to plant their vineyards. Scions that are grown in a nursery situated in a vineyard on shale soil, for example, are then grafted on to American rootstocks. The rootstocks themselves are often grown in southern Europe where soil and climatic conditions are

Figures 33–37.
Modern way
of cultivating
American rootstocks
through the year.

completely different. These two individuals are then joined or grafted to each other in a vine nursery and raised in nutrient-rich soil on a lowland plain in a site unlike a vineyard. They are given fertiliser and sprayed a lot. When these young plants are ready, they are taken and planted, for example, in a biodynamic vineyard on some extreme site above the Mosel.

For wine growers with a holistic mindset such practices make little sense – and yet it is common even for organic plant material to be produced like this. Rudolf Steiner warned: 'You can't think so mechanically about it, but you should be clear in your minds that things, by their very nature, are tied to certain localities because they have become accustomed to them. This must be taken into consideration. Otherwise you can probably achieve momentary successes, but not anything lasting.'[2] If we take this indication seriously, we need to ask ourselves as wine growers, whether it is now necessary to develop a completely new concept so that young plants can once again be produced that are adapted to their location.

Historically speaking, vine grafting has made sense in the short term having saved the European vine. But now, after a hundred years, vine development has not come very far. We are working with a very narrow range of genetic material.

To summarise, the vine we have in front of us today is an out-of-balance organism. It should therefore come as no surprise that biodynamic preparations have a particularly strong and immediate effect – their main effect is after all to bring harmony and balance. The problems associated with the grape vine, said Rudolf Steiner, 'cannot be dealt with effectively by the science available today.'[3] They can, however, be addressed if we are prepared to accept the approach drawn from anthroposophy in terms of both content and method. In his course on agriculture, Rudolf Steiner gives guidelines for this and we need to learn to understand them better. I also think that it is important with the help of this knowledge to develop further methods and, if possible, to also present them within a scientific context.

There is a long catalogue of questions linked to vine cultivation which urgently need answers. The observations made by practitioners are particularly important. Every wine grower should by rights be engaged in research and take a certain ownership of it. Wine growers also need to meet and share their experiences with each other directly. They should not simply leave the work to the institutes, but should communicate with them and provide them with practical grounding.

The impulse viticulture brings to biodynamics

In the agriculture course Rudolf Steiner speaks about the agricultural individuality. Every single farm is conceived of as a unique and self-contained individual organism. What arable and livestock farmers must first be made aware of, seems self-evident to the wine grower. Wine growers know about the special qualities of their vineyard site and how from it good-tasting, character-filled, individual wines can be produced – wine growers know about their *terroir*.

Through observations carried out over many generations, wine growers have gradually discovered which variety of vine produces the best results on this or that specific vineyard. It often needs a mere few metres or a somewhat different aspect, despite the same variety or processing techniques being used, for the wine to have a completely different quality and character. This is why in many regions a system of site classification has been introduced. One speaks of the 'first' or 'greater' site, in France there is the *Grand cru* and *Premier cru*. In the past these sites were often taxed at different rates and a 'greater' site was subject to a higher levy. The *appellation d'origine contrôlée*, the certification given to indicate certain French geographical locations for agricultural products like wine (the term means 'protected designation of origin'), came about in recognition of certain micro climatic and geological conditions.

It is obvious to most wine growers that the human being has a strong effect on the particular wine too. This is not primarily about concrete measures such as pumping, pressing or filtering, etc. Many winemakers develop a very fine and delicate relationship to their wine. There is, from the moment of harvest, a particularly intimate process of accompanying the wine and communicating with it. There are winemakers who go into the cellar each day, taste the must and then consider how best to guide the fermentation. Out of this inner communication intuitive decisions are often made. For this reason, wines also reveal a certain imprint of their maker. Engaging with the personality of a winemaker can be a very

enjoyable experience. In old cellars a taste unique to that vineyard can emerge and be passed on down through the generations. Professionals can often recognise the very specific quality of a vineyard in a blind tasting.

Wine connoisseurs are uniquely aware of special wine, special wine growers, special vineyards and of course special wine-growing areas. There is also sensitivity among consumers who enjoy good wine, for what may be termed a 'principle of individuality'. In the world of wine there is a perception of individual uniqueness. The individuality principle can be considered a fundamental feature of wine culture. In what other field of agriculture do we find such an approach to a product? Do we distinguish between carrots that are grown on sandy or limestone soils? We could of course perceive and taste these differences just as we do with wine. Growth habits are also different. Or again, how is it with milk? It was clear to many old farmers that the herbs available to cows on limestone soils are different to those growing on sandstone. This of course has a significant effect on the flavour of the milk. And yet we don't have any *Grand cru* dairies. Something comparable does however occur with cheese production. Here the principle underpinning *appellation d'origine contrôlée* is expressed and there is an awareness that fodder, regional conditions, the place of storage and those engaged in processing, have an effect on the taste of cheese.

The consumer can in this way become sensitive to the local differences in other fields of agriculture too. Most people can accept and experience that water drawn from a limestone water source has a different taste to that rising from sandstone. Strangely enough though, the huge differences in taste between spelt grown on limestone and that grown on sandstone soil, pass them by. Wine growing could in future help to stimulate this kind of awareness. The awareness living in wine production should be cultivated and made use of, yet this special asset of vine cultivation is being lost today. Nonetheless it is this particular principle that needs pursuing, developing and cultivating.

PART 3

A Viticulture that Respects the Vine's True Nature

5

THE AGRICULTURAL ORGANISM

Jean-Michel Florin

To convert a vineyard to the biodynamic approach is a huge challenge, and especially so since most vineyards today operate as monocultures with little biodiversity and no livestock. It was different before the Second World War. In Alsace, France, for example, wine producers would keep a couple of cows for milking and as working animals and therefore maintained some meadows. In the Bordeaux region it was common practice at the beginning of the last century for shepherds to graze their flocks in the vineyards during the winter and spring. The introduction of tractors and the powerful drive towards agricultural specialisation led to the current situation where no crop other than the grape vine is to be seen in the landscape.

Every wine grower responsible for guiding the journey from the soil and the vine through to the wine cellar tries as far as possible to bring the specific *terroir* quality to expression. For it is the special nature of the site as a whole – its geological foundations, the quality of the soil, its particular aspect and so on – that gives the wine its unique quality as perceived by the senses when it is tasted. In his agriculture course Rudolf Steiner said:

'The farm is always an individuality in the sense that one farm is never the same as another. The climate and soil conditions are the very foundation of a farm's individuality.'[1] This means that the growing site is not merely an interchangeable factor of production, but a significant part of the totality out of which the product arises.

To conceive of the farm as a living organism requires an inner change of approach, to switch from a thinking that is focused on factors of production (we normally refer to plant *material*, for example), to one that involves working with nature – that is with living beings. If we allow ourselves to entertain such a living and observant approach, we can gain new experiences. A wine grower from Charentes, in Western France, once said to me: 'Before I converted to the biodynamic method I never went on my fields – I sent my workers instead. But now I walk there as often as I can. I have the impression that previously my vines were like chickens in a factory farm and that now they are free and living like hens that are allowed to roam.'

It is indeed the case that the web of relationships between the vine plants and their surroundings is severely reduced in the monoculture of a conventional vineyard as a result of working with clones and seeing natural diversity as a disturbance – as competition for example, or carrying the threat of parasites.

The biodynamic approach can support the qualities of the growing site and bring them more clearly to expression. In terms of the general improvement in both the health of the vine and the quality of the vine, this is particularly noticeable when the biodynamic preparations (and especially horn silica) have been applied.

Following these initial experiences and insights, the question then arises as to how a landscape or a farm organism should be structured – how should the *terroir* be developed and cared for?

Since it is not possible to enter into a full and extensive discussion of this theme, we will limit ourselves to offering some guidelines and suggestions that will find relevance in the succeeding chapters.

Biodiversity

A first step could be to (re)create a multi-faceted and biodiverse landscape. A rich and interesting green cover can be created by using the very good seed mixes available today (see Chapter 7). Then there are the various transition zones such as access lanes and wooded boundary zones which can be redesigned. In the past, vineyards were places of great biodiversity. In Alsace many fruit trees used to surround the vineyards – in the first place the acclaimed 'vine peaches' (*pêche de vigne*), quince hedges, cherries, plums, figs, almonds and even roses, which were also grown as fungal

Figures 38 and 39. A comparison between a monotonous and a diverse landscape.

indicators. There were wild hedges containing a diverse range of shrubs. Such plant diversity attracts numerous insects and small animals; nest boxes can also be installed in the trees to encourage more birds.

Before the systematic introduction of herbicides, medicinal plants grew freely on the sunny vineyard slopes such as wild thyme, oregano, lemon balm as well as several legumes – again, all plants which attract insects.

Werner Michlits, a wine grower from Austria, demonstrates how islands of biodiversity can be created in a flat landscape using trees, shrubs and herbs. He even gave up growing vines in rows with great benefit to the landscape (see Chapter 6).

The design or renovation of a hilly or mountainous landscape with dry stone walls, stony mounds or terraces can also serve to increase the range of plant and animal species.

In order to start recreating its biodiversity we need to have a good understanding of the site. It is helpful to produce a botanical map. Older inhabitants might be asked to help with this as well as botanists, biologists and gardeners. This will ensure that locally adapted plants are chosen. It would also be worthwhile exploring the cultivation or planting of species that could balance out the one-sided nature of the vine with its vigorous climbing growth habit – this could be practical research for the future.

Animals

The second aspect of importance when designing a farm as an organism, is to develop a feed and manuring cycle. Wine growers don't usually keep animals any more. They tend to use very little manure or compost, but the presence of animals is nonetheless important.

In the first place this is because plants and animals form two parts of a whole. In the agriculture course Rudolf Steiner said, 'In the household of nature, plants give and animals take.'[2] As a being that is open towards the world the plant is complemented by the animal's internalising capacity. The animal takes in the atmosphere (for instance of the landscape) as well as the substances created by plants. It internalises or 'ensouls' them, and thereby brings them to a higher stage of existence.

Secondly, the soil also needs animal substance in order to maintain and improve its long-term fertility.[3]

The question thus arises: how can wine growers keep animals on their holding or in the vicinity? Innovative solutions are needed that often require partnership arrangements. There are some interesting attempts in this direction; we will show some examples in the next chapter.

Several wine growers are now integrating chickens among their rows of vines (see Chapter 8, 'Seresin Estate, Marlborough, New Zealand', p. 132). Mobile hen houses are ideal for this, and it has been found as a result that hen manure has a positive effect on the vines. The vine is a woody climber that human beings have transformed into a little tree. Rudolf Steiner described in the agriculture course how there is an intimate connection between the animal kingdom – and birds in particular – and trees.[4] The manure of birds applied in small amounts would therefore seem a highly suitable fertiliser for fruit trees.[5]

The new atmosphere that arises as a result of increased biodiversity and in particular the return of animals to the vineyard, brings joy and enthusiasm to the people working there. The wine grower Jean-Paul Zusslin describes the effect of sheep as follows: 'In winter our sheep are in the vineyard … but even in summer – when they are no longer there – we still sense the presence of sheep on this plot of ground even if only because here and there a few strands of wool are still hanging. The sheep are part of the vineyard even when not actually present. We have the impression that the quality of our wine has been improved by the sheep.'[6]

Enhancing the site

Through steps such as these the wine grower can form a new relationship to the landscape and their farm. They should try and grasp the unique character, the *genius loci*, of their holding ever more consciously. Against the background of this living understanding they can then implement all the practical measures. This can be assisted by the Goethean observation technique whose step-by-step process of conscious perception, makes an ever deeper and more conscious connection to the site possible.[7]

When wine growers think about their vines, it is usually the whole plantation with all its idiosyncrasies that comes to mind – not only the soil and micro-climate but also the work to be done. The name of the place often expresses the atmosphere of a plot – or at least its concrete situation.

The attempt can be made in the first place to perceive and describe the atmosphere of the place more consciously and deal with it in an individual way. It is worth taking time to walk through the vineyard on a regular basis, to sit down and use all of our senses to try to perceive its atmosphere. To retain it, we can take notes, or even try from memory to create a symbolic image. How do I perceive my plantation in terms of the elements (earth, water, air, light, warmth)? Is it dense or open, rounded or sharp, moist or dry, exposed or protected, light or dark, warm or cold and so on? Polarities of this kind can help focus our awareness on the overall qualities rather than on details. We might even try to place ourselves in the position of a plant and then ask, how do my surroundings feel?

These perceptions can be of help in getting to know the land better. They will also affect how we work because we then become more sensitive to any imbalances that occur.

A second step, which requires a bit more time and can be usefully done with a friend or colleague, consists of trying consciously to grasp the whole farm as a single entity. The creation of a kind of atmospheric map can be helpful. We can, for example, walk across the entire holding without saying a word and try and discover the qualities of the elements. Afterwards (perhaps in small groups of four or five) the qualities experienced can be drawn on a sheet of paper.

This process offers the possibility of discovering the principles for making a whole-farm design. Completely new ideas can emerge, such as using hedges or orchards or dry-stone walls. What is important here is to start from a holistic overview of the farm and not random actions.[8]

The vine is an extremely sensitive plant which reacts immediately to changes in its environment. It can therefore be described as an indicator plant that can show how the plantation is progressing. The road from a monoculture to a farm organism is a long one. It is possible though to follow it step by step. It is often necessary to work together with colleagues or neighbours and perhaps create an organism together (very much a future orientated task!). Jean-Paul Zusslin describes it as follows: 'To be a superb biodynamic wine grower on one's own is almost impossible, but when two or three people come together and complement each other's work, a great deal is possible.'[9]

Figure 40. Atmospheric mapping helps with perceiving the elemental qualities of the farm.

6

FROM GRAPEVINE MONOCULTURE TO A DIVERSE VINEYARD

Werner Michlits

We live in Burgenland in the easternmost corner of Austria on the Hungarian border. Our farm, the Meinklang Vineyard, is managed by our extended family. This brings us directly to the theme of biodiversity. We are six very different members of the family and only through the community is there an overarching commonality. Each one has their own place on the farm and, looking back, it seems that each has been able to realise their own preferences. That all three sons would remain on the farm and pursue agriculture was not something our parents had expected.

In our diverse mixed farm, which would in the past have been typical of the region, we have an arable section where we grow old cereal varieties (Emmer, Einkorn and spelt wheat). In recent years we have been working a lot with old varieties of seed. Viticulture and fruit-growing with apples and elderberries are important enterprises and since the early 2000s we have also been keeping livestock, primarily cattle. These were soon joined by pigs, horses, sheep and poultry.

Wine growing was already very far removed from the diversity of the whole farm. The last decades have also seen the introduction of more and

more machines and technical solutions. Childhood memories of the days spent in the vineyard under the cooling shade of trees, climbing in the cherry tree while our parents were working, collecting nuts in the wet grass on misty autumn days, looking for and eating the forgotten grapes left hanging on the vines after harvest, it all formed a delicate image of how diverse and interesting vineyards once were.

At the beginning of the 1990s we gained certified organic status. Ten years later the next challenge was becoming a Demeter certified biodynamic holding. That also marked the beginning of our work in keeping our own cattle. Working with animals was a major step for the family in building up the farm. Suddenly there were other living and sentient beings to care for. What do they need? What should be done to keep them satisfied?

Keeping animals on the farm meant caring in quite another way. Looking at an animal is quite different to looking at a plant. If the animal gets ill or its life comes to an end it affects human beings in quite a different way to a vineyard that falls prey to a hail storm or suffers a plant disease.

No wine grower needs to be an accomplished livestock breeder, but I do believe that keeping five sheep or ten hens is already enough to set the farm on the way to becoming a farm organism. Time takes on a new rhythmic

Figure 41. Family as part of diversity.

quality and brings a further dimension to biodiversity in the farmer's work. It is also a luxury – in the positive sense – to enjoy the raw milk of one's own sheep or cows, or the eggs from one's own chickens. With its soul the animal brings a wholeness to the landscape. It also stimulates a holistic and cyclic thinking in ourselves. How can I make use of their manure and how can I grow food for the animals?

Diversity starts with observation

Why am I writing about subjects that appear to be so far away from a bottle of wine? Biodiversity is not only about biological diversity. Biodiversity begins first and foremost with perception, in feeling, thinking and ultimately in the practical actions we take.

After all, why should technical thinking and rationalisation be more significant than the decline in the species of flora and fauna? The philosophy of biodynamic agriculture enshrines this diversity, indeed it is born and developed out of this diversity. Diversity enables living farm organisms to arise which can ultimately find their own individuality – farm organisms which are made fruitful through diversity can develop themselves further and become more fruitful.

Let us look at the definition of biodiversity given at the UN Biodiversity Convention of 1993. Here biodiversity is described as the variability existing among living organisms of whatever origin as well as the ecosystems to which they belong. A distinction is made between four levels of diversity: the genetic diversity of an individual species or the ecosystem as a whole; the number of species; the range of habitats and ecosystems; the functional biodiversity. Alpha and gamma diversity, the species' diversity of a site and the total species diversity in a landscape respectively, are the technical terms. The most well-known measure for describing diversity is the Shannon-Wiener index. It takes account of the number of times each species appears in relation to the total number of species and individuals.

We can now approach this matter from the practical side. If we look at vineyards, we generally see vine after vine – everywhere a monoculture. There isn't a single tree, bush or hedge anywhere. If I walk through a vineyard or drive past it, I can have the impression that everything here is beautifully in order. And as a farmer I might think that everything is practical and accessible, it can be worked very well by machine. The rows can be readily cultivated, driven through easily and there is space to turn around at the end. Such an arrangement also optimises the time factor: easing the workload reduces the amount of handwork needed, and also means the number of staff can be reduced.

But what of the diversity which underpins the ecosystem and brings balance, where has that gone? Already in the 1990s, we understood its value and importance. Covering the ground between the rows with green plants became a big issue at the time. Now twenty-five years later, more and more conventional wine growers are using green manure plants in their vineyards, thus attempting to bring some diversity into the vine monoculture. Vines grow in the same place for many decades and the tractor always drives along the same tracks. The soil compaction caused can be ameliorated through a well-conceived green manuring plan, the organic matter content can be increased, creating a habitat for numerous living organisms, the humus can be built up and the water holding capacity enhanced. These are only some of the best-known examples. Yet with just the vine and a number of green manure plants do we really have a diverse biotope?

Our next line of thought leads to the insects as well as fungi, yeasts and bacteria. Many years ago we set up insect hotels and put up nest boxes along with nesting places for the birds. In the end, however, these all seemed to be artificial measures for an artificial monoculture. That is how we came to seek a route away from monoculture and towards a self-sustaining and beneficial system.

Islands of diversity

I will once again take a look at the vineyard. What effect does such a vineyard have? I feel it has a very capitalistic aura. Our wine production is strongly influenced by capital and commercialism. If we listen in to the conversations among wine growers, we find that most of the time they are discussing how big the yield was, what the average price for a bottle is and how much closer the vines could be planted to get even more grapes and better quality.

While we are looking at our vineyards in this critical way, we can also reflect on the fact that a vineyard has something militaristic about it – the vines stand in long lines like soldiers and are drilled into a corset made of poles and wire. How can such a mode of cultivation – which today is accepted without question – accord with the plant's inmost nature? How can a plant find itself in such a system and live out its nature? What quality, what message can the vine impart to us via its fruits, its gifts?

On our farm in Burgenland we therefore asked ourselves how we could break up the architecture of these rigid lines of vine plants while meeting our own objective of bringing about diversity in the vineyard, all the while taking care to be practical and economically viable. We acknowledge that we live in the twenty-first century and have tractors that drive through the fields. It is not possible to convert back to horse cultivation from one day to the next. That was also an important argument for the whole family – to remain practical and take small steps forward. A few years ago we therefore started a pilot project in our largest 11 ha (27 acre) vineyard. We decided to take out some vines in order to make space for other plants and living organisms. It resulted in droplet-shaped islands being formed which still permitted us to drive along the rows with a tractor and work with implements. We laid out twenty-seven such islands. It was important for us to have many different shrubs. Many of these were dug out from overgrown windbreaks and hedges in the surroundings. Others were not native but nonetheless valuable for bees and other insects. On one such island there was a large tree, one fruit tree and various bushes and shrubs. At the end we planted cereals, squashes and other extensive vegetables that could seed themselves in this vineyard. Our aim was to eventually make these islands completely self-sustaining. They were to become a kind of biodiversity hotspot in the vineyards.

Insects and birds are quite scarce in the vineyard, but when they are present they come in droves, very deliberately and close together. Starlings for example come in a large flock, not in small groups and it feels like an invasion. They flock together and move from one haunt to the next without being connected to any particular place. Every bird needs its tree. Each tree has its own special character, its own aura and draws a particular type of bird to itself.

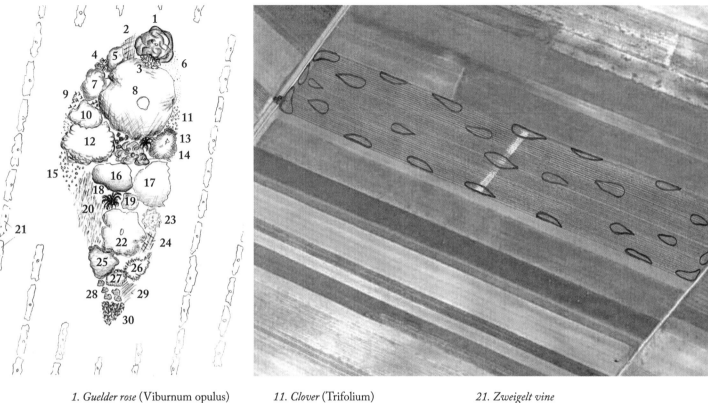

1. *Guelder rose* (Viburnum opulus)
2. *Rye* (Secale cereale)
3. *Pile of stones*
4. *Sunflower* (Helianthus annui)
5. *Blackthorn* (Prunus spinosa)
6. *Wild millet* (Panicum milleaceum)
7. *Sallow* (Salix caprea)
8. *Rowan* (Sorbus aucuparia)
9. *Maize* (Zea mays)
10. *Privet* (Ligustrum vulgare)

11. *Clover* (Trifolium)
12. *Elderberry* (Sambucus nigra)
13. *Dogwood* (Cornus sanguinea)
14. *Dog rose* (Rosa canina)
15. *Stinging nettle* (Urtica dioica)
16. *Bird cherry* (Prunus padus)
17. *Lilac* (Syringa vulgaris)
18. *Fern*
19. *Cotoneaster* (Cotoneaster divaricatus)
20. *Wild flowers*

21. *Zweigelt vine*
22. *Quince* (Cydonia oblonga)
23. *Basket willow* (Salix viminalis)
24. *Spelt* (Triticum spelta)
25. *Dwarf cherry* (Prunus fruticosa)
26. *Sandthorn* (Hippophae rhamnoides)
27. *Box* (Buxus sempervirens)
28. *Marigold* (Calendula officianalis)
29. *Einkorn* (Triticum monoccocum)
30. *Lavender* (Lavendula spica)

As above so below in the earth. Each tree has its own worms and quite specific larvae and micro-organisms. Trees and animals, the rich diversity of organisms on our 'islands', create a living organism where everything supports and carries each other. It takes a while of course before such islands are established. In our case it wasn't easy – it was too dry in the summer, then the trees froze in winter or the voles chewed the roots. We had to replant the trees three times. Many things were tried out before we discovered what really suited our site and we could create a multi-level

Figures 42 and 43. Islands of diversity in the Meinklang Vineyard. Left: sketched planting plan. Right: aerial photograph

ecosystem. What is certain however is that these islands create a different atmosphere in their surroundings.

When they find out about our project many people say, 'That will never work in our area, there is too much water stress or it is too poor in nutrients. It simply can't work.' It does work nonetheless. It is always important to allow the heart and our feelings to speak, for it is not possible to achieve everything through reason. Viticulture, which has evolved over thousands of years, has changed drastically during the last decades. So much has been industrialised and so much technology introduced that the idea of 'natural viticulture' being possible, seems far-fetched. When we come with natural intentions and ideas, they seem unthinkable and implausible. To be convinced by such 'primal information' requires a lot of effort because it has been so suppressed. It can be found however deep inside each one of us, and as soon as we are engaged with it and start following this path, we will develop an ever-stronger feeling for it.

Not pruning

In 2001 we chose a vineyard where we decided deliberately not to prune. We had come across the minimal pruning approach during a presentation at Geisenheim University. That was the first time I had come across alternative approaches to pruning. We then began asking ourselves: why do we actually prune the vine? The vine has apical dominance. It always has a leading shoot. Pruning means that this natural state is interrupted. Each time it is pruned, whether in winter or summer, the vine must readjust itself and develop a new leader.

It is also important to consider the course of the year in this connection. From the winter solstice onward, the plant directs itself towards vegetative growth. Until the summer solstice its main priority is the production of shoots. As soon as the sun reaches its highest position the plant begins to focus more and more on generative growth and invests its strength in the production of fruit and seeds. If the shoot tips are pruned during this time the vine is thrown off balance and is forced to produce a new leader. It is a measure which strongly interferes with the plant's inner order. The layman is told: the summer pruning of shoots is necessary so that the vines growing out of the stems tied to the wire do not suffer wind damage, break off and lie on top of other shoots.

Minimal pruning is in essence a mechanised management approach. After extensive investigation we came to the conclusion that minimal pruning has only found limited application. Vines managed with minimal pruning may be compared to a Thuja hedge. Using a rotary mower the row of vines is continually cut back to form a square hedge. The resulting new growth is even more intensive and the vine out of balance to a yet greater extent. Manual work is rendered unnecessary and the attempt is made to mechanise the vineyard still further. This is done a lot in Australia and harvesting is also done by machine.

At that time we were young revolutionaries and said to ourselves, 'We are going to stop pruning our vines altogether. We are just going to try it.' We called our system the 'non-pruning system' – and I consider it a harmonising method of management. We converted 1 ha (2½ acres) of vineyard to this approach. It had been included on my father's list of plots to be replanted and so was easier for us to contemplate the risk involved. We had no idea at that point what changes might result. The first two years were very unharmonious and there was a lot of rampant growth. A freshly pruned vine has a lot of power and can send out shoots two or three metres long. It is quite understandable that if its growth is continually curtailed, it will, given a chance, use all its power to grow. After a three-year transition period we found that under the non-pruning system the growth of shoots was now only 20–30 cm (8–12 in). As the years went by the growth beneath the foliage became woody and the new shoots developed on the outside. Due to the shade the outward growing shoots no longer developed buds inside. A very beneficial micro-climate with good air circulation developed on the inside and the young shoots hung loosely on the outside in full sunshine, giving a harmonious impression.

The vine has more shoots and we have many grapes, although the grapes are very small. With many small fruits there is proportionally more skin and therefore less juice. That however is precisely what we want – we want to produce extract-rich and concentrated wine and in a natural way. We didn't set up any trial plots or make comparisons, but we know from practical experience that these are always the ripest grapes on our farm. Each year we meet the Oechsle grade requirements (the hydrometer scale that measures fruit density) without any problem. There is a **high** level of overall health; the thick skins on the fruit have greater resistance too. One thing we do need to be aware of is to plant the vine **rows in the direction of the** prevailing wind, and to use strong posts.

We are very impressed and excited by our non-pruning system and see a great future in it. We have since converted three further vineyards. Work is spread extensively throughout the year but comes to a head at harvest time. To compare, under the non-pruning system we need fifty harvest hands for one day on one hectare. It often involves a very challenging search for grapes that are hidden amongst all the foliage.

For us, non-pruning is an exciting route into the future. It involves thinking wholly about the plant and its rhythms rather than placing immediate human needs in the foreground and organising everything according to practical and technical expediency.

7

UNDERSOWING VINES IN ORGANIC VITICULTURE

Matthias Wolff

Most vineyard soils today are significantly lacking in humus. Among the causes are firstly the practice of keeping the soil open – first by hand with a hoe and then mechanically with a rotavator – and secondly the consequences of land consolidation which takes no account of soil make up.

Apart from inappropriate cultivation procedures, the main reason for soil compaction is that the soil is unable to retain the finer soil particles due to its lack of humus and living plant cover. Whenever it rains these are washed down into the soil below. As the speed of absorption slows, these particles accumulate and gradually solidify in the soil.

Although similar processes occur with arable crops, the ground is cultivated (even if only lightly) between successive crops, unlike the perennial fruit and vine crops. The incorporation of crop residues leads to the accumulation of a certain amount of humus and especially to an exchange of gases – two factors of vital importance for soil fertility.

The long-held practice in viticulture, by contrast, has been to sow areas down to grass. This 'green cover' is left alone for years and at most cut as mulch. It is very rarely ploughed or cultivated.

Another significant reason for the heavy compaction of our vineyard soil is mechanisation. Economic considerations have meant that almost everything that was previously carried out by hand is now done using machines. The result is a huge number of passes with the tractor, with a corresponding effect on soil structure.

The tractor (often too powerful, too heavy, with tyres of the wrong type or with incorrect air pressure) is always forced to drive in the same tracks thanks to the generally very narrow inter-row spacings. Furthermore, plant protection measures are often undertaken when soil conditions are unfavourable, for example, when it is too wet. Bearing in mind these two factors it is not hard to imagine the condition of the soil beneath this apparently healthy, green and grassy covering. It can be confirmed through a spade diagnosis: beneath the grass, regardless of soil type, there is at most 5 cm (2 in) of loose, root-permeated soil. Below it the soil is often compacted, without root penetration, earthworm activity or any other sign indicating a living fertile soil. It is hard to see how vines can thrive under such conditions.

Figure 44. Bee-friendly flowers sown in the vineyard.

Figure 45. Vines growing on an estate in the Champagne region of France.

Pseudo solution: water-soluble fertilisers

Those observing the various fertiliser recommendations for viticulture will immediately conclude that to provide the vines with a supply of nutrients in such dead soils, the only solution is to force-feed them soluble fertilisers. This approach appears to work well on the face of it but the associated dangers are well known. Amongst other things, the nutrient-holding capacity in most of our vineyard soils is extremely low due to the lack of life and a deficiency of humus. The consequence of supplying water soluble fertilisers, especially nitrogen, is their leaching into the ground water.

If common sense prevails and less nitrogen is applied then, depending on the soil's humus content, vines will make it visibly apparent that no further transfer or uptake of nitrogen by the soil will occur, although the nitrogen may benefit the grass. The nutrients released through decomposition of green mulch will be sufficient for the grass but come too late for the vines.

The effect on the vine of such reduced fertiliser applications will manifest itself as weakened or stunted growth. Yields, as well as the quality of the grapes, will likewise decline. The resulting wine will lack substance despite the reduction in quantity. All this will be accentuated in a dry year.

This is the painful experience which all vineyards had to go through in the past if they converted to organic production but forgot to begin by converting their soils. Although much has been learnt in this regard, many managers still make this mistake and pay a high price for it when converting.

All-round soil fertility is essential to ensure harmonious vine nutrition. Without it, healthy growth and the satisfactory development of grapes for producing high-quality organic wine, will not be possible.

The challenges of green cover planting in organic viticulture

From what has already been indicated it is clear that there is more involved in sowing green manure crops than has generally been recognised. It is not enough for the plants to grow fast enough to allow passageways to be driven through as soon as possible. The seed mix needs to include plants with many different root types to encourage soil life and open up the soil. They should also include those that can penetrate deep down and break through hard pans.

Another element of this green cover should be to encourage insects. This has a particularly important effect in the context of perennial crops like the vine. Experience shows that in diverse, perennial stands of flowering plants a particular insect species rarely becomes a pest because such one-sided development is usually hindered by its natural enemies. For this, however, the plant mixture needs to contain many different flower species (*Umbelliferae*, *Compositae*, *Labiatae*, etc.) which ideally flower from

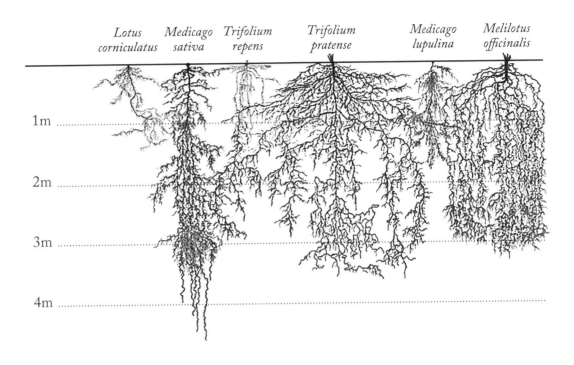

Lotus corniculatus   _Medicago sativa_   _Trifolium repens_   _Trifolium pratense_   _Medicago lupulina_   _Melilotus officinalis_

1m   2m   3m   4m

early spring to late autumn. Dead, woody plant remains that can serve as overwintering quarters for insects are also desirable.

And last but not least the green cover planting has great significance for the supply of nutrients to the vine. This is not only about the nitrogen, which legumes and other plants are able to bring into deeper levels of the soil through their symbiotic relationship with nitrogen-fixing bacteria. It is the total root mass that contributes towards the enlivening of the soil by providing nourishment for all the life in the soil. The more varied the type of root and the greater the root mass, the more diverse will the life in the soil become.

The years spent actively searching for optimum solutions in organic viticulture have shown that we cannot avoid the need to plough up and reseed at regular intervals – the sins of our fathers cannot be rectified in one fell swoop. There are also economic pressures, such as increased labour costs, that force us to rationalise our work, to 'super-mechanise'. Each time a machine replaces work in the vineyard previously done by hand, another pass is needed by the tractor on the same tracks, causing further compaction.

Figure 46. Root penetration of the soil with leguminosae.

Reseeding is necessary

Building up and maintaining soil fertility within an organic vineyard requires the regular reseeding, or at least loosening, of existing green cover plantings even if they are not yet fully mature. Of course, the frequency with which this work is carried out can be reduced as the system improves without causing stress to the vines.

Soils which are heavily compacted and where there is little or no humus take a long time to recover. To speed it up, skilful mechanical aeration can be combined with biological loosening. Mechanical aeration can be carried out for example with a mole plough, cultivator or sub-soiler. This is of course far easier to do in a fallow field than in a vineyard that is fully planted up. Biological loosening is carried out beneath the layer of mechanical aeration by plants that form a tap root while the top soil, which has been broken down mechanically, is brought to life by the fine-rooted plants.

Short-term green manure crops can have the quickest effect since the ground can be cultivated prior to sowing. Overwintering mixes (rye and vetch), perhaps with peas and/or crimson clover added, are specially recommended for this purpose. Summer green manures with a large proportion of legumes can of course also contribute towards the building up of humus and enlivening the soil.

Due to climatic conditions such as competition for water, it is not possible for such short-term, quick improvement, green manure crops to be grown in every region. It has therefore become common practice in organic viticulture to establish a rolling system of ploughing up and re-sowing every second pathway whenever the weather that year permits. It is usual to apply compost, composted manure, or other organic materials like straw, bark and/or pomace at the same time to support humus development.

Such work as spreading compost and organic materials in the vineyard, preparing seed beds and carrying out mechanical aeration are best done in late winter or early spring. At that time the soil is either frozen or dried out sufficiently to allow heavy machinery to be used without causing a lot of damage to the soil. Apart from this the soil is very cold and decomposition activity is minimal thereby reducing the danger of nutrient leaching.

A conversion mixture commonly used at present is one whose main constituents come from the Hohebuch mix (see p. 115). Perennial tap-rooted legumes are added to enrich the mix; depending on soil type these might include: lucerne, sainfoin, alsike clover, tree clover, serradella, lupin.

In order to enhance the diversity of flowers for insects, all kinds of flowering herbs are added to produce a mix of forty species. Different kinds of flower mixes are available on the market. Two examples which have proved their worth are referred to here – Bienweidemischung (bee flower mix) and Würzfuttermischung (herbal mix, see p. 115). What is important when creating such mixes is to ensure that the quantities of seed chosen for each variety are carefully calculated so that no one species comes to dominate and crowd out the others. Many organic vineyards across Europe have been working successfully with this basic mix (adjusted to soil type) for twenty-five years.

There are often problems sowing these seeds due to their different sizes, for example, between clover and peas. This is generally not a problem for seed drills used in agriculture, but in viticulture they have not yet been developed. The above-mentioned seed mix is therefore often altered by replacing large seeded peas with other legumes. It is worth mentioning that these mixtures grow and develop differently in different situations. Their diverse make-up allows each site to have its own very diverse range of root and flower qualities.

None of this is guaranteed as there are so many uncertainties. There may be wet years when it is difficult to find the right time to cultivate the ground. There may be a plague of slugs that like to devour everything that germinates. Older stands of legumes and herbs can also be thinned out by them. Or there may be dry years instead; the winter rains may be sufficient for germination to occur, but in the absence of further precipitation the young plants are in danger of drying out.

It can also be due to one's own mistakes in managing the green cover that the fine image of diversity disappears. Cutting too often and too short and within one or two seasons reduces a beautiful flower meadow to a one-sided grassland with all the problems referred to at the beginning.

In recent years repeated mowing has often been replaced by rolling in order to prevent the disappearance of species for as long as possible.

This allows many plants to continue developing till their seeds mature and enable annual plants whenever possible to self-seed.

Depending on the time of sowing, sowing depth, soil type, and also the condition of the soil at the time of sowing and later, individual species can become dominant (these are generally the annual species like phacelia, mustard, buckwheat, vetch, radish). It is therefore worth carefully observing them and, if necessary, intervene with a high-level mow. If this opportunity is missed one may well have a magnificent display of flowers – but only for one year since all the perennial legumes and herbs will suffocate beneath the leaf mass of the annuals.

In years when there is little winter rainfall and in regions with a predominantly dry growing season, the green cover plantings need carefully looking after. On soils with low rainfall it is worth planting short-term winter mixes. These are then turned in during early summer and the ground re-sown with a summer cover crop. In otherwise moister regions, where the ideal described earlier can be practised, it may be necessary to abandon the idea of having full green cover that season because of a dry winter and too little rain in the subsequent spring. It may also be necessary, for the sake of providing the vines with sufficient water and nutrients, to turn in at least every second green alley. A cultivator is often all that is needed to turn in the green manure, although it can depend on the degree of dryness. Once the water balance has been restored, green cover planting can be resumed.

Building up and maintaining soil fertility is the focus in organic viticulture. Beyond this, the aim should be to bring into and maintain as much insect diversity in the vineyard as possible in order to hinder the development of pests.

To initially improve the soil it is often wise to work with short-term cover crops, to enable the use of both mechanical and biological aeration techniques. Organic matter, preferably as stable humus, supports soil improvement.

Conversion mixes, composed of annual and perennial soil improvers and combined with a diverse range of flowering herbs, have proved their value.

To secure a successful take of seeds, good seed bed preparation as well as the right timing is important.

For managing the green cover, use of the roller is preferred to the mower. The result is a long-lasting and diverse green cover planting.

The local conditions of the site – annual climatic and soil conditions – are particularly important when considering green cover crop establishment.

Changes in viticulture in the last generation or two

- Many of the tasks previously carried out by hand have since been mechanised, from vine pruning through to the harvest.
- Repeated passes by ever heavier tractors on the same tracks (necessitated by close planting of the perennial crop) inevitably leads to soil compaction.
- The space for the vine's root development is drastically reduced.
- Soil life has become minimal.
- Inter-row green cover is usually one-sided and consists solely of grass, producing a monoculture. This does nothing to encourage biodiversity.
- In such soils the adequate provision of nutrients is only possible with soluble (nitrogen) fertiliser with all the consequences it brings for the ground water.
- The consolidation of land carried out in many areas has resulted in the remaining areas and even the plot itself suffering a significant decline in fertility.

All of these factors contribute to vines becoming more prone to disease and hence to them being exposed to ever more intensive pesticide use. This led to the widespread appearance at the beginning of the 1990s of the phenomenon of atypical aging (ATA). The last-mentioned issue on the list is of course also due to genetic impoverishment (cloning, vegetative propagation) and rootstocks that lack vigour.

The challenges of introducing green cover crops in organic systems

- ❦ Building and maintaining soil fertility.
- ❦ Quick growth to provide shade.
- ❦ Rapid root penetration to permeate with life.
- ❦ Deep rooting to enlarge the rooting zone and break through compaction.
- ❦ Rich and diverse root mass to feed the soil life.
- ❦ Build and maintain humus.
- ❦ Encourage and retain a diversity of insects.
- ❦ Wide range of different flowering species.
- ❦ Long season of flowering.
- ❦ Continuous flowering.
- ❦ Overwintering quarters for insects.
- ❦ Supply of nutrients to the vine with the help of green cover plants.

Conversion Mixtures

Flower mixes designed to provide a range of herbs, flowers and overwintering quarters for insects.

The Bienenweidemischung (bee flower mix)

Phacelia
Buckwheat
Mustard
Coriander
Marigold
Black cumin
Radish
Cornflower
Mallow
Borage
Dill
Sunflower

The Wolff-Mischung (the Wolff mix)

Alexandria clover (7.5%)
Sweet clover (7.5%)
Sainfoin (20%)
Black medick (5%)
Crimson clover (7.5%)
Lucerne (7.5%)
Persian clover (5%)
Phacelia (2.5%)
Alsike clover (2.5%)
Winter vetch (25%)
Bee flower mix (5%)
Spice mix (5%)

Mixes designed to improve the soil through deep root penetration and the activation of soil life.

The Hohebuch Gemenge (Hohebuch mix)

Feed peas (23%)
Crimson clover (10%)
Buckwheat (25%)
Alexandria clover (7%)
Summer vetch (20%)
Phacelia (3%)
Lupin (10%)
Sunflower (2%)

The Würzfuttermischung (herbal mix)

Burnet
Caraway
Ribleaf plantain
Chicory
Yarrow
Wild carrot
Parsley
Fennel
Kidney vetch
Bird's foot trefoil
Parsnip
Sweet vernal grass

BIODYNAMIC VINEYARDS AT WORK

To illustrate the variety of practice in biodynamic viticulture, we have a number of contributions from vineyards in Alsace and Bordeaux in France, from California and from New Zealand.

The Zusslin Estate, Alsace, France
Jean-Michel Florin

In 1997 Jean-Marie Zusslin converted his vineyard in the Alsace region of France to biodynamic practices. His children Jean-Paul and Marie took over the management and continued in the same direction placing particular emphasis on biodiversity and the concept of the farm organism. In the course of time various measures were put into practice which gradually created a new atmosphere in the vineyard – not only with regard to the vines but also among the people who work there. This can be perceived by visitors and extends even into how the wine tastes.

Increasing diversity

The first task concerned the natural establishment of green cover plants. After several attempts at artificially establishing green cover, Jean-Paul Zusslin decided to opt for the natural approach which is better able to respond to differing soil qualities. One section of the estate is on limestone and the other on sandstone. The identical vine variety leads to the production of wine with differing characteristics – which itself is an expression of the atmosphere and spontaneous vegetation present in the vineyard.

An opportunity arose to use a relatively protected 3½ ha (9 acre) plot of woodside land on a sandstone hillock (Liebenberg Castle) to implement measures that could further enhance the diversity of animals and plants. In one corner, close to the forest, a pear tree orchard had been planted. The flowers of this plant in the Rosacae family attract insects and serve to increase biodiversity in the vineyard. It is worth mentioning here that not very long ago, hedges of quince, hazelnuts, vineyard peaches and even small orchards, as well as semi-wild hedges with damsons, sloes and hawthorn, were commonplace in the wine-growing landscape of Alsace.

The Zusslin family also came across a shepherd who agreed to graze his flock in the enclosed 'trial' vineyard during the winter. Jean-Paul Zusslin has an interest in the animals, not only because of the manure they leave behind, but above all because of the atmosphere they bring to the vineyard, which those working in it can perceive. If any of them had spent the day in the company of the sheep they would continue speaking about it in the evening, and even in summer, when the sheep are grazing somewhere else, they still accompany people in soul and spirit.

A beekeeper also came to the vineyard and stationed his hives there. Thanks to the fruit trees, the green cover plantings and the proximity of the woodland (where many acacia trees are growing that were originally planted to produce the vine support posts), they find plenty of flowers.

A bird enthusiast created nesting places around the buildings and in the drystone walls, giving shelter to blue tits, coal tits and also hoopoes.

Importing from outside

For some time, partnerships with biodynamic livestock farmers in the nearby Vosges mountains had been developed to obtain a supply of cow manure. Tractors are equipped with small engines that require less fuel. Because they are not so heavy these tractors don't compact the soil so much – which is important for the life in the soil. Water used for making teas, preparations and the cleaning of equipment comes from a spring on one of the farmers' lands. The preparation plants are gathered on other biodynamic farms, out in the wild or in the gardens of a company producing organic cosmetics. The electricity used is 100 per cent green energy.

Draft horses

A further step was the decision to carry out some of the work using work, or draft, animals. This again required a partnership that suited the vineyard. A person trained to work with draft animals was employed and a working horse was purchased. The new employee brought a second horse with him. From then on some of the work was done using the horse, for example soil cultivation and transporting boxes of grapes. A horse does not compact the soil as strongly as a tractor, which over time produces two solid tracks between the vine rows.

The horses play an ever more important role at the heart of the vineyard as well as in the hearts of everyone involved, both great and small. Marie visits them every evening with the children. Sometimes she even collects

Figure 47. Working with horses in the vineyard.

the children from school with the horse – a special event for the village which had seen nothing like it for fifty years. There is now also the possibility of offering horse-drawn carriage rides to family, customers and friends.

Searching for partners, loving innovation

Most of the steps taken towards achieving diversity, especially the integration of livestock, was made possible by actively seeking out partnerships. It is not easy for a wine-growing operation to integrate animals. Skills must be developed and the people found who are interested in caring for the animals in the long term.

In earlier times many of the wine estates were part of a greater structural diversity because the whole family was working in agriculture from the children through to the grandparents. Each one had a task to fulfil. Nowadays we have to build living structures such as these by forming new partnerships. This requires entering into an exchange of ideas without being fearful of innovation.

Château d'Esther, St Loubès, Bordeaux, France
Jean-Michel Florin

Having worked as successful food connoisseurs in Paris, Eva and Thomas Fabian came to the Château d'Esther wine estate with its 5½ ha (14 acre) plot in St Loubès near the Dordogne, some 20 km (12 miles) from Bordeaux, where they produce biodynamic wine (*Appellations d'origine contrôlée* Bordeaux and Bordeaux Supérieur).

This place has been used for growing wine since the end of the seventeenth century. The soil is fertile and rich in alluvial sediments. There is a water course in the middle of the estate which was previously used for transporting wine.

'After we took on this estate it **was** completely restructured in 2001 as part of its conversion to organic viticulture,' said Thomas. 'It was granted organic status in 2006 and later on received biodynamic certification too.'

In the Dordogne region surrounding this vineyard there is a special

Figure 48. Vines growing near St Loubès.

climatic situation connected with the Atlantic winds that sweep in from the ocean. From the house we can see how the water level of the tidal river regularly rises and falls. The tidal flow lends this landscape a very special quality.

The Fabians were lucky. One side of the estate was bordered by an abandoned garden, the other side by the river and on the third side was a small area of woodland. This was an ideal situation for a biodynamic vineyard since chemical spray contamination from neighbouring vineyards was virtually non-existent.

The vineyard was managed organically from the beginning and the application of various plant extracts and herbal teas served to build resistance and strengthen the vines. Thomas increased the biodiversity of his holding by planting a number of plant species which he gathered himself from the surrounding area (and didn't have to buy). He took some

of the plants from the Natura 2000 wetland area that is 3 km away. There one finds stinging nettle, comfrey and ferns. Other plants like St John's wort and field horsetail grow in the immediate neighbourhood. Herbs like garlic, sage and chives were grown on site.

Another important aspect of biodiversity management on the wine estate is that of maintaining and increasing the diversity of animals. In cooperation with the LPO (France's bird conservation society) Thomas began installing a large number of nest boxes on poles at the entrance to each vine row. There are at present around 100 nest boxes at Château d'Esther of which 70 per cent are occupied.

Nest boxes for bats were also made and put in place. Bats are insect-eating mammals and valuable predators of the moths considered vineyard pests such as cutworms and sawflies.

To encourage insects and small creatures, various berry-bearing shrubs and fruit trees (such as peach, pear, apple, apricot, nectarine and various plums) were planted between the vine rows or along the borders. Because the fruit trees currently demand a lot of attention in terms of pruning, the sale of fruit is being considered to supplement income in the short term.

Since 2014 Thomas has one cow on site so as to have an 'animal presence' and manure available for compost making. He purchased this organically raised animal from a stockman friend to graze the grass between the rows in winter and early spring. This was a new challenge which meant building a cowshed and having someone available throughout the year to care for it.

Figure 49. The wine cellar in Château d'Esther.

Eco Terreno, Sonoma County, California, USA
Robert Izzo and Daphne Amory

In 1980, Mark Lyon and his father purchased Eco Terreno, an existing vineyard located on the Russian River in Alexander Valley in Northern Sonoma County, California. The purchase was purely as an investment property as Mark was employed as primary winemaker at a major Sonoma winery. The vineyard had been planted with vines from border to border and was farmed along the conventional practices of industrial vineyard farming. Encompassing 50 ha (120 acres) of flat, rocky riverbed made up of Cortina and Yolo loam, the rootstock consisted of St George field-grafted to Cabernet Sauvignon, and of AxR1 grafted to Sauvignon Blanc. Mark and his father were most interested in the wine-tasting expression of true Bordeaux varietals, and found this site was able to capture those without being in the Napa Valley.

Over time nematode pressure began to undermine the original rootstock choices and the first generation of new inter-plantings began, incorporating different rootstocks while still field-grafting varietals to create a diversified blend for winemaking. Cabernet Franc, Merlot, Petit Verdot, Malbec, Semillon, Chardonnay and their varying clonal selections were introduced. The grapes continued to be sold at a premium price to other wineries while new buyers were brought in for the introduced clones. Always the focus was on economics, yields, market share, and, most importantly, critics' ratings of finished wine. The vineyard was economically viable, though costs continued to escalate. The monoculture, dependent on synthetic inputs and heavy cultivation, produced a bleak landscape with pressure from disease building. It was becoming unsustainable.

As Mark travelled throughout the various wine regions, he began to form a different picture of how he might farm wine-grapes. While in Bordeaux in the early 2010s he tasted wines that carried a strong expression of place, and he discovered that they were farmed biodynamically. Returning home to his vineyards, Mark noticed an absence of sound and of soil cover that had been present in France. He saw a bird on the ground searching for food while an employee

Figure 50. Cover crop at Eco Terrino.

was spraying a pesticide nearby, dressed in a white hazmat suit. This observation sparked a farming change – one that would dramatically shift the landscape of all 60 ha (149 acres) under Mark's stewardship.

He hired a management team to help transition the property to biodynamic practices, and brought in a consultant to oversee and facilitate the change. Within two years the original landscape of exposed soil, drip-fed nutrients, and the fullest complement of herbicide and pesticide sprays, shifted to one of full cover crops, no drip-fed inputs, nor any use of pesticides. Within five years the landscape had been transformed, with over forty-five differing and interdependent grass, legume, and flowering species – under vine and in row – and large wildlife and bird populations. The continual incorporation of manure-based compost, biodynamic preparations, compost and plant-based tea sprays, minimal tillage and

cultivation practices began to be noticeable in improved vine health and wine quality.

Human interaction with the vineyard was also changing. Moving from an outside management-based operation, Mark hired an internal team that operated from the vineyard proper in partnership with a new management team that had permanent staff working on site with the internal team. This began his personal move away from his full-time winemaker role. He recognised the purpose and potential of Eco Terreno as a viable brand and organisation to implement the threefolding aspects of biodynamics thoroughly throughout its business model.

Since 2016, Eco Terreno has incorporated their farming practices into the understanding of business modelling and human capacity engagement for the further enrichment of social growth. The on-site team has been tasked with integrating the principles of a diverse working biodynamic farm into the existing landscape of a typical, Californian monocultural vineyard footprint. Beginning with the earlier work of cover crop diversity based on soil type and nutrient builds, the gardens, orchards and olive groves have evolved with the understanding of not only feeding the team at Eco Terreno, but of creating differing enterprises that support community engagement. Chickens have been

Figure 51. Bee garden at Eco Terreno.

integrated into the vineyard as working partners, managing the soil floor and insect pressures while supplying eggs and meat. Goats are integrated for riparian and inner vineyard corridor management. Their role is one of aiding air and light flow, which helps to alleviate insect and mildew pressures, manages invasive species, and strengthens the river corridor.

The on-going integration of diverse systems and animals has shifted the understanding of the vineyard team from one of singular focus to a more complex understanding and questioning of how best to support the systems that are present, while always building the potential future of the wine farm. Recognising that people are an integral component, interwoven with the land and essential for developing the collective wisdom of the larger farm unit, Eco Terreno has committed to providing full-time employment, a living wage and comprehensive benefits to every team member in the company.

As Eco Terreno grows in the marketplace, so does the community scope. The purchase of a building in San Francisco to house the tasting room and restaurant of Eco Terreno has increased the diversity in the farming systems to support the city venture. The city allows for a collaborative relationship to develop between urban and rural understandings of farm and place associated with regional food systems and the integrated manner in which they develop. Community continues to develop and stories become the vehicle for symbiotic growth.

Bringing organically and biodynamically grown produce into San Francisco also requires the establishment of a new type of transportation and distribution system, allowing for a design based on cooperation rather than capitalism. Shared operational elements such as reduced carbon emissions, flexible scheduling, alternative advertising and a co-op style budget account for the unique challenges of growing produce based on social and environmental principles.

In understanding the build of soil relationship from the early days of transitioning, Mark has clearly defined the web of connectivity that continues to expand diversity within the farm, the marketplace and the culture within the Eco Terreno team. This growing recognition of integrated systems continues to inform the business culture of Eco Terreno as it explores new ways to communicate the lessons being learned through the practices it has incorporated.

Littorai Wines, Sonoma County, California, USA
Ted Lemon

Littorai Wines is a family estate located in western Sonoma County, California, and dedicated to vineyard-designated Pinot Noir and Chardonnay wines. Littorai began its biodynamic journey in 2000. After a twenty-year career of conventional farming in Burgundy, Oregon and California, my wife Heidi and I, as proprietors, had become dissatisfied with this farming paradigm as an explanation for how plants grow and reproduce. Today Littorai itself farms half of its total production, the remainder is from growers. All Littorai's vineyard sources are farmed using organically certified materials, 85 per cent are farmed biodynamically. There are fourteen vineyard sites spread over 100 km (65 miles) from north to south.

The fertility cycle is focused at two estate properties, one each on the Sonoma Coast and Anderson Valley. We produce our own compost for all the sites that we farm. For the grape growers with whom we work, we provide preparations and compost as requested by the individual growers. Some follow their own biodynamic paths and others follow the Littorai program completely. The compost stream is composed of as varied materials as possible, sourced from our properties: grape pomace, hay balage, tree coppicing, vineyard prunings, cow manure and summer crops of sorghum and sunflowers. Supplemental hay and cow manure are sourced from organic dairies and farms as needed.

The animal world is an important part of Littorai's practices. Dexter cattle, primitive Shetland sheep, ducks, chickens and a donkey create diverse manures for both open pastures and vineyards blocks. The Dexter and Shetland breeds are small-statured, hardy, thrifty, self-reliant animals that produce healthy offspring without human intervention.

Littorai grows all the plants required for its biodynamic teas and preparations and makes all its preparations in-house. Most of the animal parts required for the preparations now come from the Littorai animals.

For many years we have concentrated on the development of healthy, complex farm organisms for the two primary sites. Part of this development is an extensive tree planting program. In a summer dry climate, trees are a critical, drought resistant, self-regenerating source of food (and shade)

Figure 52 (left). Coastal fog above Littorai Vineyard, Sonoma, California.

Figure 53 (right). Building a layered compost pile at Littorai. After layers of grape pomace, vine prunings and cow manure, a layer of hay is applied.

for the animals, compost materials, **and biodiversity**. On the primary sites, vines represent less than 16 per cent of the total property acreage, the balance is forests, fields and hedgerows. Redwoods, native deciduous and live oaks, poplars, native hawthorns, native cedars and a variety of willows are the backbones of this program. Additionally, we have established companion plantings of perennial and annual shrubs and flowers at all of the sites we farm. In the early years, these plantings were primarily composed of European species, well-adapted to our environment (for example, lavender and rosemary). In the mid-2000s Littorai established a trial garden for other species. Since that time, we have shifted largely to plantings of native plants.

Insect diversity and habitat are critical. After many years of managing European honeybee hives, we abandoned the effort. We now rely on wild honeybee colonies, which live in abundance in our area, to handle the ecological niche which the European honeybee has occupied for centuries

in North America. The more beautiful question to us was, what native pollinators performed the services of the honeybee before its arrival in the Americas? The answer is, of course, a wide array of species, but important among these are native American bees including *Osmia californica*, *Osmia lignaria* and Leaf-cutter bees. We now have many constructed nests for these species, and have installed bat and owl boxes as well as bluebird and swallow nests throughout the properties.

Vineyards pose unique challenges as biodynamic 'farms' in that the demands of wine production easily compromise farm diversity. Indeed, most of the traditional winegrowing regions in both the New and Old Worlds are vast monocultures. We find it hard to understand how the biodynamic model can thrive under such conditions over the long term. Our goal is to contribute to a new wine-farm model, where wine production can remain the central focus, but additional farming activities perform the vital role of nurturing the long-term health of the vines, our consumers and the farmers.

While wine from biodynamic grapes and/or biodynamic wine certainly qualify as food products, the alcohol content means that our wines present health risks that other biodynamic produce does not. For this reason, and as a contribution to the greater community good, Littorai grows biodynamic vegetables for a local food bank, harvesting tons of produce each year. We intend to grow this program in the future and involve other local biodynamic vineyards in delivering these healthy vegetables to those most in need within our community.

In our present age we humans are fully responsible for the future of our planet. Solutions to our problems will come from human action in concert with nature. The challenge before us is as simple as it is complex: can we imagine an ecologically *generative* agriculture? Are humans capable of creating a beneficial synergy with nature, leading to results which nature alone cannot achieve? Can farms generate ecological health? At Littorai our driving force is not biodynamics as we have inherited it from the twentieth century, but a reinvigorated biodynamic paradigm of generative, alchemical, ecological agriculture. As one author wrote recently, 'Integrating human work into an ecological vision has a broader potential, which is to reconfigure the relationship between the natural world and the human economy.'[1]

Figure 54. Sheep at Littorai Vineyard.

The new viticulture will be a viticulture *of* the environment, of the fields and streams and forests, of the vines in communication with the cover crops and the companion plantings. Systemic resilience, immunological response capacity, dynamic equilibrium, carbon sequestration, water table recharge, species diversity, all these will become part of our everyday viticultural language.

At Littorai these ideas inspire us. Lili and Eugene Kolisko called their core book on biodynamic farming *The Agriculture of Tomorrow*. Our commitment, and our challenge, is to create the wine farm of tomorrow.

Seresin Estate, Marlborough, New Zealand
Colin Ross

Seresin Estate is a wine estate in New Zealand, owned by the Hollywood film maker Michael Seresin. His wish was to produce a first-class wine, of correspondingly high market value, using a production approach that has a social, spiritual and ecological orientation. What is unique is the varied way in which animals are included in the management of the vineyard. This requires a suitable structure.

The vineyards have passageways of varying widths – 2 m, 2½ m, 3 m (6½, 8, 10 ft) and differing cordon arrangements (the wires are set at heights of 80 or 100 cm, 30 or 40 in). An irrigation hose is attached at 50 cm (20 in). The fruiting zone is at around 1–1⅓ m (3–4 ft). The vines are between ten and twenty years old. Almost all the vines on the estate are trained as cordons and pruned once a year. The posts used for training the lines of cordons have a diameter of 10–12½ cm (4–5 in) and are stable enough to withstand the pressure of large animals like cows and horses leaning against them. The sheep can pass beneath the lines while the cows and horses remain in the passageways. There is either permanent grass growing in the passageways or a carefully chosen sequence of arable cropping. There are also poultry which are moved around the estate in mobile units.

Pasture and feeding

Areas for grazing outside the vineyards were created and hay is made on the non-cultivated areas. This is used to provide feed for the animals in the period when the vines are growing.

The grass ley is made up of a mixture of clover, plantains, chicory and grasses. It is only mown once it has flowered and produced seed and, in order to encourage self-seeding, is often only partially mown. What is cut is then used to create a long-lasting mulch beneath the vines.

Sheep

Once the grape harvest is complete the sheep are brought in to graze the vineyard intensively section by section. The ground beneath the vines is then cleaned up before growth starts in spring. The sheep are removed from the vineyard before the buds start to burst and prior to cultivation. For every five hectares there are about 300 lambs (24 lambs per acre) which means there is about two weeks of winter grazing on each place. The sheep are kept on a set area using an electric fence. When the grapes have grown to about pea size, around 50 sheep per ha (20 per acre) are allowed in to graze off the leaves. It is important here to keep an eye on the sheep and move them on in good time so that they don't cause damage. They must be out of the vineyard at the very latest by the time the grapes start colouring. If everything goes to plan the grapes will be

Figure 55. Biodynamic viticulture on the Seresin Estate, New Zealand.

exposed and not be damaged. The height of the cordons must be set so that the sheep can only remove the leaves to the desired level.

We have built up a flock of around 100 sheep with our own rams. Our sheep are of the Wiltshire breed, which are very resilient and do not need shearing. In addition, we bring in 100 Merino lambs from a neighbouring farm in winter for grazing.

Cows

Cows can graze most of the year in certain parts of the vineyard. We chose to have Jersey cows and have a milk and dairy processing unit close to one of the vineyards. It is an area of activity requiring a lot of work and ongoing attention. The passageways are 'strip-grazed' in winter. A mobile water trough and electric fencing are used for this. One part of the electric fence is permanently set up next to the vines. Each year we have a cow in milk within a group of four animals of varying ages. During the main growing season of the vine the cows graze in the olive groves that are trickle irrigated. The dung from the lactating cows is used for making compost. During summer, cows and sheep graze on the same field.

Horses

In 2008 horses were brought into the vineyards and olive groves of the Seresin Estate in order to spray the preparations. In the space of three years we developed, together with another farm, a horse-drawn sprayer. We then invested in some heavy horses that could pull the sprayer which weighed 1000 kg (1 ton) when fully laden. This enables us to spray 30 ha (75 acres) of land over fifteen days. Depending on the season, the spray wagon is used to apply preparations, compost teas and other extracts. In recent years many of our vineyards have not required any applications of sulfur.

The growth of grass and fencing are always important issues to be aware of. The vineyards that are grazed are surrounded with solid fences.

Poultry and other birds

Poultry and other birds are distributed widely across Seresin. The corridors where native wild flowers are allowed to grow provide habitats for many birds. For an orchard this can always have two sides but our feeling is that the advantages outweigh any potential harm. We are convinced that the birds are 'ploughing the air'. Each day there are falcons, water birds and many other species to be seen. Only very few native birds cause harm to the vines, and the great diversity keeps their populations in check. It is nonetheless necessary to protect parts of the vineyard temporarily with a net.

Poultry serve to meet many of our needs including having our own eggshells for the barrel preparation and some composts. We distribute the eggs among the staff who then bring back the eggshells. We set up mobile henhouses in the vineyards, olive groves, vegetable gardens and pastures. The cockerels are raised and fattened separately. We buy older breeds of hen chosen for their character (instead of the usual chickens bred solely for laying). However, when they ripen, the grapes do need protecting from the poultry. The high protein feed required by these creatures is difficult to produce ourselves and so we buy it in. We are in the process of establishing our own worm farm and also feed offal collected from the local butcher.

Figure 56. Using a horse to apply the preparations.

Pigs

Pigs too are involved in the production and processing of vegetables. They are very effective in working the soil and our continuing process of diversification enabled us to start keeping these animals. In order to keep pigs in the passageways of the vineyard we had to use a low-level line of barbed wire or electric fence. It is important for pigs to have places where they can wallow as well as shade and fresh drinking water. It was possible to combine vegetable production and wine growing in a couple of vineyards, a good alternative to vine monoculture. We are also considering the integration of nut and fruit trees in the vineyard for the pigs.

Diversity

The two basic management concerns regarding the keeping of livestock are diversity and simplicity. We are convinced that diversity is an important element in a self-contained agricultural system. Sustainability is the key to a stable culture. It is the global vision of the owner Michael Seresin that enables us to work with this very personal form of 'agri-culture'. The integration of animals within the agricultural landscape has been very enriching for us and is the foundation of every sustainable system of agriculture.

PART 4

How to Make the Vine Stronger

9

MANURE

Jean-Michel Florin

Any book about biodynamic wine production cannot leave out the importance of manure. This theme is so wide-ranging, however, that a detailed presentation would go beyond the remit of this book. I will therefore limit myself to an introductory sketch. Suggestions for further reading are at the end of the book. The main focus of what follows will be on the biodynamic spray and compost preparations, and therefore on the heart of biodynamics.

Let us consider the agricultural wine enterprise as an 'agricultural organism' in the way it is understood within biodynamic agriculture as a whole.

This organism extends out in two directions: firstly, horizontally in terms of the space it occupies. This is the most visible dimension, the landscape itself, in which the wine grower seeks to create a living diversity of topography, vegetation and animal life (see Chapter 5). Then there is the often-disregarded vertical dimension. This dimension extends deep into the earth where the plants' roots break down the bedrock and help convert it into living soil; with its shoot and flowering capacity it also extends in the opposite direction up through the atmosphere towards the starry cosmos.

Between these two zones there is a thin layer of topsoil in which the plant covering develops. This ground cover, which is far from being hermetically sealed, is a living skin that is influenced both from above by the atmosphere and the cosmos, and from below by the bedrock. Rudolf Steiner equated the vertical dimension of a farm organism to a human being standing on their head.[1] This comparison may seem surprising but it makes us aware of the fact that the soil is not there primarily to provide the plants with nutrients but rather to give impulses. In reality most of the biomass is drawn from water and carbon dioxide while the nutrients taken in from the soil come in very small amounts.

The vine plant grows by sending its roots deep into the earth to find the substances it needs. In this way the roots correspond in their function to the nervous system of the human being.

With the help of mycorrhizal fungi which surround the plant roots with their fine threads, the vine engages in an intimate exploration of the soil. It also exudes large amounts of organic substance through its roots that provides food for the mycorrhizal fungi and numerous other micro-organisms living in the soil. The vine is thus actively seeking the nutrients it needs on the one hand and, on the other, contributing directly towards the vitalisation of 'its' soil.[2]

The vine also produces a shoot that grows in the air and light, develops leaves and later transforms them into flowers. At the end of the growing season the entire mass of organic material produced by the plant (except the seeds and wood), dies back and falls to the ground to be broken down and converted into humus.

Biodynamic fertilisation

As we have already seen, the manure needs to come primarily from the farm itself (see Chapter 4). Rudolf Steiner explained in his agriculture course that 'A healthy farm would be one that could produce everything it needs from within itself.'[3]

According to Rudolf Steiner, the aim of biodynamic fertilisation is primarily to keep the soil as healthy and alive as possible by imbuing it not only with substances but above all with forces. An understanding of these forces is needed in order to understand the role played by the preparations.

Manuring or fertilisation occurs concretely on several levels:

- On a *physical-mineral* level using appropriate soil cultivation techniques to stimulate soil processes.
- On the level of *plant growth* to bring the soil back into balance and re-enliven it through spontaneous plant cover or green manuring – in effect to lift it into the plant kingdom or kingdom of life. In biodynamic viticulture plants are sown, some of which are native and others specially chosen to counteract the imbalances caused by growing the vine (see Chapter 7). The various plant species release different substances through their roots and thereby enrich the soil. These secretions can also release nutrients and minerals from the geological substrata. Wherever possible, all plant residues are composted using the biodynamic herb-based preparations and returned to the earth.
- The incorporation of an *animal element* is important for both plant health and the production of humus, and this applies to the vine as well. That is why some manure is mixed in. It doesn't have to be much but should be of high quality. It is generally cow manure that has been carefully composted using the six compost preparations. An important goal of composting (including the composting of manure) is to create humus, but this compost can also be spread among the vines for its fertilising effect, especially in springtime. In order to have the effect of prepared compost during the conversion period or on steep slopes, the compost can be applied in liquid form.
- A fourth and far more subtle level of manuring is provided by the *biodynamic preparations* which bring forces to the soil and build a bridge to the cosmos. It is particularly on this level that the attitude of the individual farmer or wine grower becomes important. Rudolf Steiner also emphasised how important it is to develop a personal relationship to the manure. This requires an active engagement with all the elements and qualities of the farm. Personal connections to the plants and animals will then be strengthened.

The biodynamic preparations

Biodynamic compost – and ultimately all biodynamic manure – is characterised by the fact that is treated with six fermented medicinal herbs. These preparations, used in minute doses, may be compared to a sourdough starter which guides the ripening process in the compost and manure so as to produce the best possible humus. They don't work in a direct causal fashion like artificial fertilisers, but rather serve to engender cyclical processes and have an all-round regulative quality. 'The effectiveness of the preparations may be understood as aiding the individualising tendency within the life of nature by giving it structure and stability.'[4]

A path towards understanding the six compost preparations may also be found by considering the plant species used as medicines for the earth. All of them can grow in challenging or one-sided situations which they harmonise and bring into balance through their mere presence.[5]

Compost is a kind of all-round medicine, a preventative that can address the many imbalances occurring in the soil and the climate, something that is of particular importance today with the challenges posed by the climate crisis.

According to Rudolf Steiner the preparations also stimulate chemical processes in the soil and plants.[6] This doesn't mean they simply produce chemical substances, but rather that they make them available to the crops and regulate their effects. It is about the living chemical processes occurring in nature, and not the dead chemical reactions that take place in the test tube.

An overview of the preparation plants

In his course on agriculture, Rudolf Steiner characterises the nature of the preparations in a brief and poignant way.[7] In doing so he also describes the active processes of various chemical elements like sulfur, potassium, calcium, iron, etc. and how they relate to the preparation plants.

Yarrow (*Achillea millefolium*) grows on compacted, warm and dry soils. It forms a ramified rhizome beneath the soil which helps to maintain the layer of humus. The plant develops a hard stem with very finely divided, dark green leaves that culminates in in a dense, whitish flower head. As a medicinal plant its healing properties in relation to the kidney and bladder system is well known. Trials have shown that in situations of potassium deficiency the yarrow preparation can help unlock the potassium present in the soil.[8]

Chamomile (*Matricaria recutita*) grows on soils that can be very compacted and mineralised, and re-enlivens them with its thick rooting system and rapid, vigorous growth. As a medicinal plant it is used primarily to treat intestinal complaints. Its calcium-related activity counteracts similar conditions in the plant.

Stinging nettle (*Urtica dioica*), which grows on piles of decomposing plant-refuse or waste tips, has as its objective the goal to 'make the manure intelligent' as well as the soil subsequently treated with it. As Rudolf Steiner expressed it, 'Adding *Urtica dioica* in this form is really like an infusion of intelligence for the soil.'[9] As a plant it is capable of transforming a waste tip into fertile soil. It regulates the soil's iron processes and can have a positive effect on the photosynthesis of cultivated crops.

Of all the deciduous trees of Europe, the oak is the species which can develop the strongest vegetative power. It is noted for its longevity and its innate capacity to resist the processes of decay. Even when there are many dead branches, it is time and again able to produce new shoots. It continually seeks a balance between over-vigorous vegetative forces and the excessive drive to form wood. **Oak bark** (especially that of the common oak, *Quercus robur*) gathers a lot of calcium as it grows. As a result, it has an ameliorating effect on the vegetative process and prevents diseases which can occur on cultivated plants due to imbalanced growth.

Dandelion (*Taraxicum officianale*) is usually found growing on rich pastures containing an excessive amount of organic matter of animal origin. Its particular capacity to absorb light and cosmic influences is demonstrated by its strongly structured leaves, sun-like flowers and fine seeds. This is what enables the dandelion to regulate the surplus fertilisation effect of organic, terrestrial matter on plants and make them receptive to light influences.

The final preparation is **valerian** (*Valeriana officianalis*), a large flowering plant that grows in moist places and releases a delicate scent. It manages to survive cold and damp periods in this situation by creating an atmosphere of warmth around itself. A tincture made from its roots is a well-known sleep medicine. It is used in biodynamic agriculture because of its relationship to the phosphorous process and its 'warming' capacity.

This is only a brief sketch. To gain a deeper understanding of these plants there would need to be an in-depth exploration of each one in much the same way as was carried out in Chapter 1 for the vine (this is done, for example, in *Extraordinary Plant Qualities* by Bockemühl and Järvinen).

The spray preparations

Once prepared compost and/or manure has been created to lay the foundations of a healthy soil in which plants can thrive, a second step can be taken to support the plants as they develop skywards from the soil. This requires two further preparations that are applied as sprays: horn manure and horn silica. They serve to harmonise and strengthen the connection between the plant (vine) and its surroundings.

These two preparations – to the great surprise of many – require cow horns. To gain an understanding for this, we need to know that the cow, so fundamental to the agricultural organism, is the animal with the most highly developed metabolism. Cows are experts in digestion and devote a large proportion of their lives to it. That is why their manure has a quite different quality to that of other animals – it is truly a concentrated life force.[10] Their horns furthermore have a special relationship to the animal's metabolic processes. The horn manure preparation is made out of manure which has been put into cow horns.

In order to enhance and make the forces contained within the manure more potent, the filled cow horns are laid in the ground over winter to ferment. When the finished preparation is dug out of the ground in spring it smells and looks like good, colloidal, woodland humus.

The other preparation, horn silica, is quartz meal which has spent the summer under the earth inside a cow horn.

The ways these two preparations work can be understood by reflecting on the way a plant grows. In the Goetheanistic study of the vine described in Chapter 1, two fundamental principles of growth are distinguished:

- ❦ The *vegetative* stage. This is expressed in the formation of leaves and stem and all the parts of the plant where photosynthesis takes place; root development is also particularly strong during this stage. The horn manure preparation primarily supports and regulates this first phase of growth.
- ❦ The *generative* stage. This is expressed in the flowering of the plant, in the differentiation and emergence of aromatic properties. This stage leads ultimately to the formation of fruit. The horn silica preparation supports the flowering and fruiting process.

Horn manure preparation is applied when planting, as the buds open, and during the vegetative phase. It is best applied in the evening at a time when forces are being concentrated in the soil.

Horn silica by contrast is best sprayed out early in the morning in order to connect the plant more strongly with the light. It makes most sense to spray this preparation before and after flowering and up until the fruits ripen – at the time when the shoots, for instance, take on form and grow upwards. The development of the leaf epidermis, the sugar **content** and flavour of the fruit is influenced by it.

Both preparations are applied in minute quantities after having been stirred vigorously one way and then the other in lukewarm water for an hour.

In addition to compost, many wine growers also apply teas made from the preparation herbs in order to balance out irregularities in the microclimate. Where weather conditions are unusually cold, an application of Valerian extract can have a 'warming' effect, it can even be used following a hail storm. Yarrow tea applied during periods of heat and drought on the other hand can be refreshing.[11]

Why are such preparations applied?

Rudolf Steiner recommended these preparations in order to give forces to the earth and plants. All forms of agriculture ultimately exploit the soil, removing substances and forces from nature. The preparations

are applied primarily to replenish these lost forces. In the historical development of agriculture, they constitute a genuine innovation, of value to the future of the earth.

The farmer or wine grower can use these measures to enhance humus development and create living soils of such special quality that they can really heal the earth. Long-term trials of over thirty years in Switzerland and similar trials in Germany have shown that the application of biodynamic preparations enhances soil life and the development of humus. By increasing this layer of humus more carbon is also being sequestered. The widespread application of biodynamic methods could therefore also contribute to a reduction in the threat of global warming.[12]

Figure 57. Vigorous stirring of biodynamic preparations.

10

ESCA: A GRAPEVINE TRUNK DISEASE

Florian Bassini[1]

Esca is a disease caused not solely by the mere presence of a pathogen in the environment. There are many other factors that cause an outbreak. The power of the pathogen as well as the weakness of the organism affected play a role, as do environmental and climatic changes and the harmful effects of technology.

A first step towards understanding why a disease occurs (and also why it doesn't), involves knowing the history of the pathogen and how it arises in as great a detail as possible. The following questions can offer helpful pointers:

- How does the disease behave?
- Does it affect all the organisms or only a few individuals?
- Why do certain individuals not get ill or show no symptoms?
- On a general level, what does this disease wish to show us?

Degenerative diseases cause premature decay in a plant, attacking the perennial plant organs. Esca, eutypa and tortrix moth, as well as grapevine yellows are such diseases.

In the case of esca (a grapevine trunk disease), it is a complex degenerative vine disease which has increased dramatically across the world over the last three decades. The term esca comes from Latin and means 'tinder', in reference to the dry, tinder-like consistency of the wood in the later stages of the disease. It is caused by various wood-decomposing fungi. Since the 1990s esca has been spreading through the northern wine-growing regions and is particularly problematic in that it can lead to the death of the vine rootstocks.

It appears that wood diseases have long been present in viticulture. The ancient Greeks, for instance, were aware of esca and there are modern studies of esca dating back to 1865. Incidents of esca appeared to increase rapidly in the nineteenth century as the vineyards recovered from the grape louse plague.[2]

For a long time, esca was thought to be an illness of old age that to begin with affected only a few plants. For several years, however, it has also been affecting young vines of around ten years old and the proportion has been steadily increasing, and not only because treatment with sodium arsenite has been prohibited.

Such changes in the behaviour of diseases often go hand in hand with other changes. Human interventions that aim to increase yields, cultivation techniques, selection and mechanisation are likely to prepare the way for such diseases.

The most significant changes are:

- The grafting of vines at the beginning of the twentieth century.
- Clonal selection which from around 1960 displaced mass selection.
- Increased mechanisation and deeper soil cultivation from the mid-twentieth century.
- Introduction from 1976 of the standardised omega grafting technique (with hand grafting becoming solely a passion of the dedicated), which, because it was easily mechanised, resulted in many poorly carried out grafts.
- The disease appeared in a different form in the various regions according to soil and climate conditions and of course the particular variety of vine.

A research project was undertaken in 2003 and 2004 in three large wine-growing regions (Alsace, Burgundy and Provence) on 37 plots across 24 wine estates, all of which were organic or biodynamic. We assessed the degree of disease affliction on each plot (light, medium, heavy) based on the proportion of diseased vines and the speed with which the disease spread. Using a range of criteria (such as age, variety of rootstock, vine variety, yield, time of pruning, and disease survival rate) to compare the various stages gave us a better understanding of esca.

A comparison between the level of infection and age of the vine showed clearly that vineyards planted before 1960 generally had a lower level of infection, while vines planted later, and especially those after 1976, had a medium to high level of infection.

What was also apparent was that although mass selection techniques did not prevent infection, cloned plant material was significantly more susceptible to esca.

The predisposition for infection was likewise significantly lower in whip grafted vines than those grafted using the omega technique.

The relationship between the infection rate and yield of the vine is also interesting. The yield partly reflects the practices of the individual wine grower. During the last 150 years – a relatively short period of time – yields have increased significantly. This demonstrates that although lower yields don't guarantee a lack of problems, higher yields seem to render the vine more susceptible to esca.

The study documents the many factors that cause esca. This leads to questions regarding a holistic and respectful way of working with the vine. What system of management – from the choice of rootstock to the growing method – will strengthen the vine's immune system and make it more resistant to the various attacks? Furthermore, can the disease really be described as harmful, or is it rather a normal reaction to a system that is no longer in balance?

Chapter 17 (Regenerative Grafting) offers one possible treatment for vines affected by esca.

11

FUNGAL DISEASES

Jean-Michel Florin

Before thinking about how best to control one or another fungal disease it is important to first understand the role played by fungi in nature and the conditions in which they appear (this of course applies to all so-called pests). They should be seen as indicators since they point towards an imbalance between the plant and its environment. A plant never really gets ill in the way animals and human beings do. The plant continually tries to regain a balanced relationship with its surroundings and is generally able to grow new organs.

It is of course necessary here to make a distinction between wild and cultivated plants. The cultivated plant has been changed by human beings in order to produce a greater yield of fruit – the grape vine is a good example. Cultivated plants are therefore often dependent on a very specific growing environment. The relationship of the cultivated plant to its surroundings can be upset far more easily than that of the wild plant to its environment.

The French scientist Francis Chaboussou indicated a number of factors in the vineyard that would appear to disturb the balance:

- Pesticides which upset the physiological balance of the plant (this is one of his great discoveries).
- External influences (milieu and cultivation method).
- Internal factors (in the case of the vine, primarily the influence of the rootstock).

153

Chaboussou built on the observations made by Perier de la Bathie who noticed a massive increase in the presence of grey mould following the regeneration of the wine estates with grafted rootstocks. Chaboussou asked himself 'whether some of the rootstocks, though resistant to grape louse, could in fact make the plants more susceptible to mildew.'[1]

The French botanist Lucien Daniel was an acclaimed specialist in everything to do with grafting. In 1900 he wrote, 'In viticulture there are deleterious as well as beneficial grafts.' This led him to conclude that one cannot simply graft every European grape variety on to all rootstocks, since this would lead to an imbalance.[2] Lucien Daniel's observations demonstrated that the repopulation of vine plantations with American rootstocks encouraged fungal disease. The American rootstocks are stronger than the European vine varieties. This then is one cause of weakness in the European vine.

The plant archetype

The plant in its vegetative growth is truly plant. Through the process of photosynthesis it produces organic mass, whilst through its roots it tends towards the mineral and becomes woody. In the upward direction its vegetative growth is arrested and comes to a halt in the flower, which can be conceived of as a kind of parasite on the green plant: the flowering process arrests growth, the sugars in the leaves are transformed into secondary substances, and a flower, an organ composed of numerous metamorphosed leaves, is formed that generally has a very precise geometric structure.

Let us compare the fungus to this plant archetype. Beneath the earth's surface we find extensive mycelium, a kind of enormous root – some types of mycelium extend across very large areas (in Oregon for example, mycorrhiza has been found that extends across 9½ sq km, 3¾ sq miles).[3] The fungus lives under the earth or in organic materials that are either alive or decomposing. Through its mycelium it obtains the nutrients it needs. It is of course unable to engage in photosynthesis. What appears briefly above the earth and is eagerly gathered at certain times, is called the carpophore or fruiting body. These mushrooms have many different shapes, colours, smells and tastes.

When compared with a green plant a fungus appears as a kind of flowering fruit growing directly from the root (mycelium) without any vegetative part. Seen in this way, the appearance of fungal growth on a host plant can be understood as a premature or delayed flower or fruit. There are a number of phenomena associated with fungal diseases which back up this contention – the yellow or orange dust of the rust fungi, or the changes in the growth of *Euphorbia* that is caused by fungi.

Goethe wrote of the rust fungus: 'What irregularity of growth has brought the plant to the position where instead of producing offspring in a joyful and lively way, it lingers at a lower stage and completes a corrupted act of pollination?'[4]

Under what conditions does this 'displaced flowering' occur? Either when the plant grows too strongly and produces more raw plant juice than it can process through photosynthesis, or in times of drought when growth is arrested and becomes irregular. Fungal growth occurs in situations where the activity of life is insufficiently taken up by the organism, where there is too much or too little life.

In order to recognise and treat fungal diseases in good time, Rudolf Steiner recommended amongst other things a tea (or decoction) made from horsetail (*Equisetum arvense*).

Figure 58. Botrytis cinerea *(botrytis bunch rot or grey mould) on Riesling grapes.*

Field horsetail (*E. arvense*)

The first impression we might have of a horsetail plant on a sunny morning could be of a plant made of wire, a kind of plant skeleton hanging from the sky with its many threads, like a vegetative crystal reduced to its spatial-geometric structure.

In order to understand the nature of horsetail better it is worth getting to know its ancestry. Horsetails, ferns and mosses belong to the group of plants known as pteridophytes, which are counted amongst the earliest that grew upon the earth. They were once very widespread and huge in size. The primeval ancestors of our modern horsetail (calamites) grew 30 m (100 ft) high during the Carboniferous Period. That is the period when the vast deposits of coal were formed which we have been so avidly consuming in recent decades. The huge horsetail plants grew in very moist, swampy and foggy surroundings, from which they absorbed carbon dioxide and released oxygen in return. Along with mosses these plants began to

internalise their watery environment through fluid processes. It became 'channelled' and, as the sap fluids flowed through their cells, the plants began to take control of the water. These are the first vascular cryptogamae, plants without flowers and a hidden reproduction.

Where horsetail (E. arvense) grows

As the name suggests, field horsetail (*E. arvense*) grows in fields and in any areas exposed to the sun, such as on hillsides. And yet, even if the surroundings and the soil appear superficially dry in comparison to the marshy areas favoured by most of the related species (giant horsetail, marsh horsetail, etc.), its presence always indicates a compact, waterlogged layer, down below where it extends its finely divided and horizontally growing rhizome roots. Rhizomes 500 m (1600 ft) in length have been found. Some of the marsh horsetails (*E. palustre*) in Finland have been found to have a rhizome mass one hundred times larger than the mass appearing above ground. The horsetail can therefore be imagined as being a huge plant lying down, with its main shoot – the horizontally spreading underground rhizome – extending out beneath the earth and the side shoots growing up towards the light – both the brown fruiting stems and the green sterile stems.

Horsetail is thus a plant which exists to a great extent beneath the earth, and this explains why it is not easy to eradicate once it has taken hold unless soil compaction and waterlogging are addressed.

Horsetail (E. arvense) through the year

The field horsetail (*E. arvense*) is a perennial plant whose rhizomes are winter-hardy but whose stems die off in winter. In spring, brown stems appear on the banks and ditch sides which look like parasitic plants or toadstools. These are the fruit-bearing stems (sporophores) with cone-shaped heads and covered with hexagonal plates upon which the spores are found. These spores fall on the moist ground where they germinate and develop into tiny male or female prothallium. Fertilisation, which is not dissimilar to that of fungi and lower organisms, takes place in this shady-moist environment.[5]

Figures 59 and 60. Field horsetail (Equisetum arvense). Left: vegetative stage. Right: the sporophores.

The brown prothalliums soon wilt. Later, the same underground rhizomes send up tiny green (sterile) stems shaped like tiny conifers with branches that gradually expand into space. This is the best-known stage. Starting off soft and pliable, they become firmer and more brittle as the season proceeds. The leaf whorls are reduced to small brown scales and the green side shoots give the plant a very stable structure, which is further reinforced by the pronounced furrows along the stem. The plant is rougher to the touch as the season continues.

During the course of the year, horsetail (*E. arvense*) fixes a large amount of silica which it extracts from the water taken up from the soil. It has a very high rate of transpiration which can sometimes be seen in the form of tiny water droplets on the stem tips early in the morning. The silica accumulates in the epidermic cells of the stem and in the swellings that appear on the outside in the form of glass-like opals. Horsetail (*E. arvense*) thereby becomes a veritable silica castle as Rudolf Steiner described it when he recommended using horsetail (*E. arvense*) to ward off fungal disease: 'Silica lives ... as if in a castle, as in the horsetail plant, it (the silica) lives.'[6] Within the stem it is hollow and the entire plant is brittle and pliable. It is only the turgidity and stiffness caused by the silica that keeps the stem walls upright. No sooner is it picked than it collapses and loses its firmness. The horsetail (*E. arvense*) dies back in the autumn and a new cycle begins the following year.

If we compare the annual cycle of the field horsetail (*E. arvense*) with other related species, we can see how well organised the vegetative and generative processes are, both spatially and temporally:

- 🌱 The generative process of the sporophore takes place in early spring in an environment that is still cool and moist. Fertilisation occurs on the soil in the shade and damp.
- 🌱 The vegetative process takes place by the stem growing ever higher towards the light throughout the summer and transpiring large amounts of water from the ground.

The substances of horsetail (E. arvense) – *sulfur and silica*

Field horsetail (*E. arvense*) has a special affinity with two substances: silica, which can be felt in the roughness of its surface (horsetail (*E. arvense*) used to be used as a brush for cleaning pans), and sulfur, which can be found by chemical analysis.

All horsetails contain a large amount of silica, a substance with a strong connection to the light much like the transparent clarity of a rock crystal, which is used in pulverised form for making the biodynamic horn silica preparation.

Although compared with other European species, the field horsetail (*E. arvense*) has the smallest proportion of silica, only 67 per cent of the ash compared with 91 per cent in rough horsetail or scouring rush, *Equisetum hyemale*; it also has the highest proportion of sulfur at 4.2 per cent while scouring rush has only 0.3 per cent.[7]

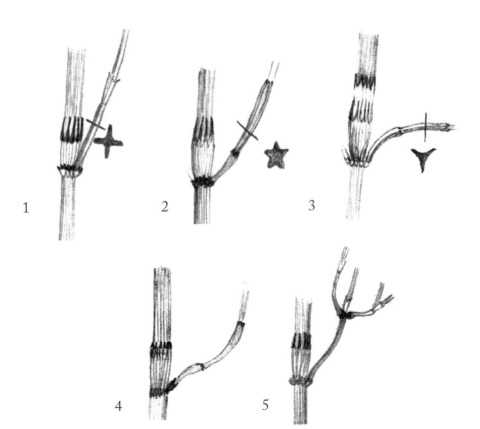

Figure 61. Distinguishing the most important horsetail species 1. Equisetum arvense, *field horsetail; 2.* E. palustre, *marsh horsetail; 3.* E. pratense, *meadow horsetail; 4.* E. fluviatile, *water horsetail; 5.* E. sylvaticum, *wood horsetail.*

It is worth remembering that sulfur is commonly used to combat fungal diseases and in horsetail (*E. arvense*) we find a special combination of two substances to regulate fungal afflictions: silica, which facilitates a good light penetration of the external cells, and sulfur which strengthens internal vigour.

> We need to relieve the soil of the excessive lunar force; we need to find some way of reducing the water's mediating capacity, of giving the soil more earthiness so that the water that is present does not absorb the excess lunar influence. We accomplish this – though outwardly everything remains the same – by making a fairly concentrated tea out of *Equisetum arvense,* which we then dilute and use as a kind of liquid manure on the fields where we want to combat blight and similar plant diseases. Once again, very small amounts, a kind of homeopathic application, will be sufficient.[8]

The gesture of horsetail (E. arvense)

Let us summarise what has already been discussed. Horsetail (*E. arvense*) is a plant of great vitality which it is able to channel and control. In evolutionary terms it is one of the first plants with a system to manage the flow of fluids. It thrives in a watery milieu in places where there is underground waterlogging and is able to regulate this dammed up water and excess moisture in the soil.

At the same time, it is a plant with a great affinity for light. We can think of the geometry of its form but also its capacity to fix silica. In view of this we could consider the whole plant as being a 'crystalline' flower, particularly since it is also able to fix considerable amounts of sulfur.

In an exemplary way, the field horsetail (*E. arvense*) separates the generative process, which belongs to the elements of earth and water and therefore more to darkness, from the vegetative process, which can be linked to the elements of air and warmth.

Understanding the effect of horsetail (E. arvense)

From a Goethean point of view it is not only the substances in a medicinal plant that are important. Even more important is the process which has led to the formation of these substances. To understand the effect of horsetail (*E. arvense*) we need a thorough knowledge of how it grows. In other words, horsetail (*E. arvense*) is not put to use simply because it contains silica and sulfur. There are many other plants that also contain silica and yet don't display the same effects as field horsetail (*E. arvense*). It is the entire gesture of the plant which makes it effective. That is why the complete plant is used and never an abstract or isolated principle.

Field horsetail (*E. arvense*) works in nature like a kind of kidney – it draws out excess and stagnant water from the ground that would otherwise serve to encourage fungal moulds. Put another way it is able to change the soil and make it less attractive to such fungi. It also has the ability to clearly separate the generative (moist, dark) from the vegetative (light, airy) process.

Apart from this, it develops the outer structure and firm epidermic cells in an extreme way during its vegetative growth, unlike other more sensitive plants which are less able to build resistance to fungal disease and are therefore more susceptible. The impression is given that the flowering process, which usually signals the end of growth in other plants, expresses itself as a formative, guiding gesture from the beginning.

In biodynamic viticulture horsetail (*E. arvense*) is used as a prophylactic measure against fungal disease. A brew is made (the plant material is left to simmer for at least 40 minutes in order to release the silica) and the liquid is sprayed out on the ground in spring. This brew can also be added to the sulfur and copper treatments during the period of vegetative growth. The addition of horsetail or stinging nettle decoction can often significantly reduce the amount of copper required.[9]

Willow

François Bouchet, a French advisor for biodynamic agriculture, introduced the use of willow bark for the treatment of fungal disease. It is an interesting example of how, through an understanding of the disease and careful observation of plants from a biodynamic standpoint, an intuition leading to the discovery of a new natural treatment can occur.

We will only sketch a brief, general picture. A full botanical study will of course take far more precise observations if only because *Salix* is a very complex plant family comprising numerous hybrids. The descriptions should serve to stimulate the reader to carry out their own further studies.

The family is made up of two different groups of species:

- Willows with broad leaves (like *Salix caprea*). These are bush willows with green-coloured and rounded leaves which are sometimes even crumpled.
- Willows with narrow leaves (like *Salix viminalis*). These are trees and bushes with a much lighter colour and many soft branches.

We will concentrate on the willows with narrow leaves which have long been used as a medicinal plant for its fever-reducing properties. Aspirin is surely one of the best-known synthetic derivatives of the original, natural willow bark substances.

The willow habitat

The first impression of a pollarded willow in a dark winter landscape is of the many shoots springing out of the trunk. This image can be compared with the flower-like quality expressed in the virginal bark that ranges from soft yellow, brown-orange, through to purple. Willows are actually shrubs since they continually send out new shoots while the older branches are regularly discarded. The older branches rapidly grow brittle and fragile – the precise opposite to the oak.

In early spring the willow forms light, silvery-gold catkins. The shoots, which are very thin and bendable, produce very narrow-shaped leaves which are frequently silver on the underside and a strong, gleaming green on the upper surface. Virtually no other tree has such narrows leaves, even the buds from which the leaves develop are comparatively narrow. What a different picture it presents to that of the alder which is often found in the same area and yet has broad round leaves.

Quite early on in the year (from May to June), the female flowers of the willow produce light and fluffy seeds that are carried by the wind and can cover the ground in a delicate blanket.

The narrow-leaved willows are pioneer plants along riverbanks; they are among the first to cover the very damp soil. They bring their glowing colours into these moist surroundings. They often grow with their roots in the water, cleansing the water. Thanks to the high transpiration rate of the willow leaves water is also given over to the air. In this way the entire wetlands are cleansed and the soil stabilised.

Figures 62 and 63 (left). Sallow (Salix caprea) *as the representative of the group of willows with rounded dark green leaves.*

Figures 64–66 (above). Left and centre: basket willow (Salix viminalis). *Right: white willow* (Salix alba). *Both are representatives of willows with narrow and pointed leaves.*

The gesture of the willow

Narrow-leaved willows possess many of the qualities of young plants:

- 🐛 Flexibility of the shoots.
- 🐛 Soft wood which decays rapidly (in contrast to oak which keeps for a long time).
- 🐛 Continuous dropping of limbs as a way of overcoming ageing and death.
- 🐛 Quick and seasonally limited seed germination.
- 🐛 They reveal an intimate connection to light in the way the leaves reflect light and the twigs glisten in winter.

In summary, the gesture of narrow-leaved willows reveals vigorous growth that is further enhanced by the moist situation. They continually produce new shoots and discard older material. The coloration of the bark demonstrates that the flowering process enters right into the mineral realm. This is where the strong connection with light is also revealed.[10]

Salicin

The bark and leaves of willow have long been used for pain relief and for their fever-reducing effect, especially with malaria. It is also one of the original sources of the modern **Aspirin** – salicylic acid. Salicylic acid is also use to combat fungus diseases which, like malaria, tend to arise under warm, moist conditions. The constituents of a plant reflect and are the result of the plant's entire biography, never the product of a chemical factory. Instead of extracting the active substance of salicylic acid and applying it in an isolated form or even synthesising it, a tea or decoction made from the bark is used in the biodynamic approach.

The willows producing the most salicin in their bark are: purple osier (*Salix purpurea*) with 3–8.5 per cent, white willow (*Salix alba*) with 0.5–1 per cent, and basket willow (*Salix viminalis*) with up to 1 per cent salicin.

Recent studies confirm that salicylic acid strengthens the systemic resistance of plants. What has been discovered by observing the processes of growth and its gestures has been confirmed by current research results.

In biodynamic viticulture the willow is also employed to counteract fungal diseases, especially mildew. The bark from young one-year shoots, ideally collected in spring, is heated in water to 80°C (175°F) – it should not boil as that would destroy the salicylic acid. The brew is then applied, often with the addition of sulfur and copper.

Horsetail and willow

Both horsetail and willow are used to counteract fungal diseases. Both plants grow in a very moist environment and both seek to connect with the light and the dry air as intensely as possible. Excess fluids need regulating in order to prevent over-vigorous growth. What can be perceived as the properties of the plant in its environment can be found in the effect of its extracted substances.

The stinging nettle (*Urtica dioica*)

In late summer our gaze is drawn towards a clump of stinging nettles growing beside the path against a wall. The ruby-coloured stems gleam upright in the morning sun with regularly spaced fresh green leaves growing out horizontally. It is a picture of structure, order and vitality in what seems at first sight to be a somewhat disordered and off-putting profusion of growth.

When we touch the nettle, tiny hair-like needles on the leaves inject us with histamine and other chemicals. We feel a burning sensation and our skin reacts with a rash. This makes us immediately aware of the burning aspect of the plant.

The stinging nettle's environment

The stinging nettle is very widespread, at least in warmer regions. It is often sufficient to take a walk around the garden or visit a site where building materials, rubbish or stones are lying around and then find a fine clump of nettles concealing the rubbish.

The stinging nettle grows from spring until late in the autumn and always seems to be green – as if it were permanently springtime. It grows with increased vigour once it has been mown. It is usually found on roadsides, along walls in the half shade and in cooler, almost damp situations. It appears wherever there is fresh organic material – manure, forgotten compost, piles of leaves, but also stones and old iron.

Stinging nettles always grow in dense patches leaving little room for other plants to grow – they occupy the entire space. The few plants that do accompany them are, like the nettle itself, nitrogen-hungry plants such as cleavers, docks, brambles and bindweed. Among all these plants that display very little structure, the stinging nettle stands there like a soldier with its upright stem and opposite pairs of horizontal leaves.

It could be said that the stinging nettle prefers places where the soil is not yet homogenous and penetrated by life, but rather the opposite: an accumulation of many different substances.

If a stinging nettle plant is pulled out of the ground, we find a vigorous yellow rhizome – an underground stem from which grow the many stems making up a clump. From these rhizomes many small roots grow out,

Figure 67. The leaf layout of the stinging nettle is strictly cruciform.

intensively penetrating the organic or waste materials. Whilst the nettle does not grow down far into the earth (it remains in the material that needs transforming), the underground rhizomes have enormous growing power and continually send up new plant stems.

The annual cycle of the stinging nettle

Early in the year the young stinging nettle shoots emerge from the earth and grow upwards without forming a rosette. The plant grows upright immediately. The angular and grooved stems develop the typical heart-shaped, serrated leaves from the beginning. Held by the striking and very rigid veins, every leaf grows out horizontally into space in a precisely determined way. With its evenly structured inter-nodes and cruciform leaf layout it presents a very structured characteristic picture.

Leaf metamorphosis is limited. The phase of contraction consists only of a reduction in the size of the leaf to the leaf point.

The leaves are remarkably well formed. There is a fine balance between the mass and substance-forming forces, in which larger formless leaves dominate, and structural forces as expressed in the feathery leaves of say the wild carrot.

Supple to begin with, the stems become harder and more fibrous after flowering in June.

The stinging nettle is dioecious which means the plants are either male or female. The flowers appear in the leaf axles of both and are unassuming, limited to the reproductive organs and short green sepals without any flower colour at all. The flower stems, reminiscent of hazel catkins, also have more of a vegetative characteristic. If we stay awhile beside a group of nettles we might see a cloud of yellow pollen appear. The closed flower sepals of the male plants open up briefly to release their pollen into the air – quite a surprising phenomenon. We could say the stinging nettle – instead of having coloured blossoms – flowers by spreading substance into its surroundings. It is somewhat similar to the injection of burning substance via the leaf hairs into those that approach the plant too closely.[11]

Constituents of the stinging nettle

There is an interesting polarity in the stinging nettle between the substances of the leaf, which has a high nitrogen content, and the fibres produced in the stem that were once used for making rope. The stinging nettle is a plant containing one of the highest quantities of plant protein (recognisable by the foul smell of badly fermented nettle manure). In addition, it forms substances similar to insect venom in its hairs with a high acetylcholine and histamine content. The stinging nettle also has the highest iron content of any medicinal plant, although this varies during the year and depending on the growing situation (it is usually highest in spring and autumn).[12] According to Rudolf Steiner the stinging nettle can regulate the effect of iron in the soil.[13]

The gesture of the stinging nettle

Through its almost military-looking structure of its leafy stems, the stinging nettle reveals its capacity to bring order into places that are chaotic. It carries this gesture into its underground rhizomes too and

Figure 68. The sequence of leaves of the stinging nettle shows only a limited development in their form.

brings order to the processes of life. It is one of those plants that can rapidly convert organic material into a black, humus-like substance.[14] It shows its great vegetative power in the fact that it grows almost continually throughout the whole year.

Its vigorous growing propensity shows itself in the denseness of nettle clumps and the leaves' capacity for photosynthesis. More than parsley and second only to curly kale, the stinging nettle contains the highest amount of chlorophyll.

	chlorophyll-*a*	chlorophyll-*b*
Curly kale	189 mg	41 mg
Stinging nettle	185 mg	173 mg
Parsley	157 mg	55 mg
Spinach	95 mg	20 mg
Broccoli	26 mg	6 mg
Green beans	12 mg	4 mg
Green peas	10 mg	2 mg
Cucumbers	6 mg	2 mg
Kiwis	1.7 mg	0.4 mg
White cabbage	0.3–1 mg	0.1–0.2 mg

*Table 1. The chlorophyll-*a *and chlorophyll-*b *content in vegetables and fruit per 100g fresh material.*[15]

The true flowering process takes place in the periphery of the plant: in the release of pollen, in the finely serrated structure of the leaves and their well-marked veins, in the development of stinging hairs (which contain the unique venom), and throughout the green parts of the whole plant. Even in the striking yellow colour of the rhizomes a kind of blossoming can be sensed which brings the light down to earth.

To summarise we can say that the stinging nettle presents a very strong and vigorous vegetative growth process – visible through its high protein, nitrogen and chlorophyll content – which gives it a very youthful quality, with a high vitality and the capacity to produce a lot of living green matter. This vegetative process is then continually met with a specialising formative process from above, as is shown by the regular leaf serrations.

This process does not weaken the vegetative growth (which it would do if hard thorns were formed or there **was a** yellowing of the leaves) but rather it *accompanies* the growth.

May this brief summary stimulate readers to experience the process described for themselves, in order then to internalise the 'gesture' of the stinging nettle, and use it with understanding.

Application in viticulture

The stinging nettle is a plant of great importance in biodynamic agriculture which, according to Rudolf Steiner, cannot be replaced in its effects by any other plant. It is one of six plants added to compost in the form of preparations. In viticulture it is used in the form of a tea or liquid manure.

Just as the horsetail is a typical stem plant, the stinging nettle is a perfect leaf plant. Through the control of its growth processes, which never grow rampant yet always remain **vital**, the stinging nettle sets a kind of example for the way leaves should grow in the vineyard, where vines have a tendency to become over-exuberant and grow rampant. It is in this context that they use their special capacity **for r**egulating the iron-related processes in the soil. It is therefore applied whenever there is a need to strengthen and guide the processes of photosynthesis, to stimulate plant growth or treat chlorosis.

Figure 69. The stinging nettle rhizome with its yellow colouring and finely structured root hairs bring a quality of light into the soil.

12

THE NATURE OF SULFUR

Marc Follmer

In the agriculture course Rudolf Steiner spoke about substances and their salts in a way that had never been done before. 'Modern chemistry talks only about the corpses of the substances, not about the real substances. We have to get to know their living, sensitive aspect.'[1]

He did not merely describe them as chemical elements defined by their specific properties (as had been done since Antoine Lavoisier and is still done today), but shed light on the activity, processes and gestures of the substances – especially when it came to characterising the role each one plays in the living world. A new way of approaching the realm of substances had been born.

In doing so, Steiner described the activity of the substances in an often very imaginative way. This opened the door to a simpler and more holistic approach which is, however, no less real. The gesture that determines each substance allows for and can indeed integrate the purely chemical properties of the substances concerned. Such an approach to substance has its origin in the Middle Ages (or even earlier) when plants were known by their common names and expressed something of their inner nature, as with the homeopathic doctrine of signatures.

The results of modern science do not stand in contradiction but can be added seamlessly to the approach suggested by Rudolf Steiner. This makes it possible to perceive the properties of substances not only in the test tube but out in nature as well – in the soil, plants, animals and human beings.

171

By using this approach, the substances and also the chemical elements can be conceived of as a kind of terrestrial anchor for the supersensory processes of nature without which all life on earth would be impossible. By studying the activities and properties of the various substances we can gradually discover their true character. By going beyond their physical and chemical properties we learn about the role each substance plays in the world and in the natural relationships of plants, animals and human beings.

In what follows we will use this approach to consider sulfur, a substance that is used widely in the growing and processing of wine.

Sulfur as a chemical element

Sulfur is positioned in the same column of the periodic table as oxygen with which it shares several properties. It is a multivalent element occurring in large quantities. In water and most other polar solvents (consisting of molecules with slight electrical charge) it is insoluble, but does dissolve in non-polar solvents. Its density and hardness are very low and it has a low melting point (115°C, 239°F). Its boiling point is 445°C (832°F).

It is distinguished by the particular yellow colour of its natural state. The main characteristic of sulfur relates to its connection with heat. As a solid it is not very fixed and needs only a little energy to change its state. It is enough, for example, to gently heat sulfur in a flame to make it lose its form and become fluid.

These properties are easily altered. There is a whole range of different sulfur molecules and – uniquely among substances – sulfur is very mobile. It is difficult to force it into a defined form.

The chemistry of sulfur is such that it can form compounds with all other chemical elements except the inert gases. It also combines with itself and appears in the solid state, as well as in liquid and gaseous states in several allotropic forms (the ability of a chemical element to exist in two or more different forms). The most stable and prevalent form of sulfur is α-sulfur with the formula S_8.

Sulfur readily combines with oxygen. More than 30 compounds have been identified, the most common being sulfur dioxide (SO_2), which is produced by burning sulfur. This is highly soluble in water where it produces sulfite ions (SO_3^{2-}). Combined with hydrogen it forms a gas, hydrogen sulfide (H_2S). In nature this comes about through volcanic activity and bacterial processes. Sulfur-carbon compounds are also found in oil and gas deposits. When combined with oxygen as a sulfate, it leaves the organic realm and can then be reabsorbed by plants which, with the sun's energy, combine it with hydrogen (another element connected to warmth) to synthesise sulfur-rich amino acids, for instance, cysteine.

Secretion by   ⟵   Sulfur   ⟶   Connecting to
living realm -SO_4 living realm -SH

Sulfur and life. (After Wolff, Grundlagen)

Sulfur in nature

Sulfur is found everywhere – in the air, in the sea, in sulfurous springs, and in the form of salts (sodium sulfate, potassium sulfate). It is found combined with calcium in the form of gypsum which slowly dissolves to release sulfur.

The most stable sulfur compounds are those combined with metals in the form of sulfides and sulfates. As soon as the ores meet the air however the sulfur is released and replaced with oxides and carbonates.

When it becomes gaseous in the context of volcanism, sulfur reveals its true nature – the existing order is shattered through volcanic processes and a mighty chaos ensues.

Sulfur is also present in all living organisms. Plants take up sulfide ions and are the only organisms able to convert SO_4^{2-} ions from the soil into sulfide compounds. Animals then take from the plants these sulfur compounds which they need for their metabolic processes.

In the soil, sulfur appears in the form of sulfate ions formed through the oxidisation of proteins in the organic waste of living organisms or their corpses. It is also formed by the bacterial oxidation of mineral sulfides.

Plants are especially rich in sulfur compounds. The sulfur-containing glycosides of the *Cruciferae* (or *Brassicaceae*) family (mustard, cabbage), deserve particular mention. The *Alliums* (garlic, onion, ramsons) also contain sulfur compounds. These plant families are notable for their great vitality and rich sap content, but they also defy a fixed structure.

Sulfur also plays an irreplaceable role in animal and human life. Sulfur – like carbon, hydrogen, oxygen, nitrogen and phosphorous – is present in every living cell and is even found in the smallest sub-particle (cell nucleus, protoplasm). Sulfur is a formative substance par excellence.

The biological role of sulfur is connected primarily with the form and structure of protein. These compounds, and indeed all sulfur compounds, are weak and very easily dissolved. This results in the formation of various other sulfur compounds. When these compounds breakdown, sulfur-hydrogen compounds are formed, producing the well-known rotten egg smell.

Sulfur, like phosphorous and all essential as well as fatty oils, is a substance that is closely connected with warmth. Warmth is the element that mediates between the non-physical and physical worlds, hence the reason why it is found in the living worlds of plants, animals and humans. With human beings, warmth is the bridge between the body and the life of soul – for instance, our capacity for enthusiasm engenders warmth in both body and soul. In nature it is warmth which brings movement to all that is static – ice melts and water evaporates, leaving ideal conditions for plants to grow.

Just like warmth, sulfur resists the tendency to harden and be bounded. In order for the spirit to become active, the forces of form and structure must be softened. This is why sulfur activity is often associated with a certain amount of chaos.

In our own organism sulfur is primarily active in the metabolic and limb system. This is where the most proteins arise and are found.

The fiery quality of sulfur is made use of in viticulture to combat mildew for example, or as a dioxide in winemaking. It is very important to understand the way it works in the context of biodynamic agriculture and its own inner dynamic. Sulfur is found in certain enzymes that are present in all living cells. It is interesting that certain bacteria use hydrogen sulfide (H_2S) like water (H_2O) in a process not dissimilar to a primitive form of photosynthesis.

Sulfur and the three alchemical processes

The alchemists considered the world of matter in terms of three dynamic processes relating to the principles of Salt, Mercury and Sulphur.*

* To distinguish the modern chemical material element sulfur from the alchemical principle or process, we spell the alchemical principle as Sulphur.

The Salt principle operates primarily among solid substances (NaCl, KCl, SiO$_2$). Here the crystalline form is dominant and the so-called 'imponderable' elements of heat and light are excluded. The Sulphur principle is found in substances that contain these imponderable elements, for example sulfur and phosphorous. The Mercury principle is the mediating, balancing element between the Salt and Sulphur principles and expresses itself mainly through the metals.

Imponderable		Measurable
Light, heat, blazing a trail	←——————→	Measurable qualities and contained density
Light		Darkness
Cosmic	←——————→	*Terrestrial*

Sulphur process: heat	Mercury process	Salt process: crystallisation
Sulfur formation. Internalisation of the 'immeasurable' in substances to allow entry to cosmic forces	Transition and the quest for balance between the forces of dissolution and densification, of release and flowing together. Droplet metamorphosis	Salt formation. Crystallisation of measurable matter, open to the influence of terrestrial forces and gravity
Form is given up. Direction given by darkness and light	Breathing rhythm. Alternate holding and releasing. Continual change of form	Creating a structure by condensing. Separation of 'light' and 'darker' matter
Day extends across the earth	Morning and evening	Night extends
Tropics. The earth incorporates cosmic influences	Temperate zones. Play of water between ice and steam	Polar regions. The earth glows above the ice
Summer. Marshy lowlands	Spring and autumn	High mountains, winter
Indentation of blossoms (especially around the petals)	Skin/surface formation (fruit formation)	Root, stem, seed formation
The scent of plants	Plant rhythms arise	Taste. Tendency to lignify
Essential oils	Saponin	Alkaline tannins, for example

Table 2. The three alchemical principles of Sulphur, Mercury and Salt. (From Vademecum, 2008)

We find these three principles at work in all the processes of nature. They form an integral part of the entire structure in the substances of plants and animals and the human being as a whole, right down to the tiniest unit of the cell.

If we wish to consider the chemical elements from the point of view of process, we cannot but apply the *tria principia*, the three primes, of the old alchemists. It enables us to understand how substances work and, above all, how they work within living organisms. Each substance of course has its own relationship to these three principles, but they provide a real key towards understanding the phenomena of life.

The name Sulphur, or *solferus*, means something like 'sun bearer'. The activity of the sun is after all borne on light and warmth. Phosphorous and sulfur carry this activity within them, or to be precise, light is given off by phosphorous when it burns in the air, and sulfur bears heat through the dissolution of form and structure. In the realm of substance these two activities are kept apart. A closer look at these two will make their complementary properties visible.

The effect of heat on sulfur leads to a dissolving of form, bringing it into chaos and even destroying solid structures. We only need to consider volcanic processes or the keratolytic (softening of the skin) effect of ointments containing sulfur. Sulfur also has a very strong presence in the human metabolism.

Sulfur is therefore distinguished by two key qualities: its connection to warmth, which, unlike any other element, is expressed in its capacity to internalise warmth on the one hand whilst on the other to bring about a degree of chaos by dissolving structures as a precondition for becoming active.

Warmth is a force in nature that mediates between the spiritual and physical worlds. Sulfur is the substance capable of mediating this warmth in the realm of life (the plant, animal and human kingdoms). As the old saying goes, 'Matter is never without spirit, spirit is never without matter.' Perhaps this gives us a better understanding of what Rudolf Steiner meant when he said:

> Sulfur is the element in protein that plays the role of mediator
> between the physical in the world and the omnipresent spirit with

its formative power. You might even say that anyone wanting to follow the trail the spirit leaves in the material world must follow the activity of sulfur. Its activity is not as apparent as that of the other elements, but that's exactly why it is so extremely important. The spirit working into the natural world follows the paths of sulfur. Sulfur is actually the carrier of the spirit.[2]

Sulfur in oenology

Sulfur is mainly used in the fermenting of wine in the form of sulfur dioxide, as it is in food processing.

Sulfur dioxide occurs as a gas under normal room conditions. It is produced by burning native sulfur in the air and is soluble. It is a form of sulfur activated by oxygen that releases heat. It irritates and affects the digestion. Those who are sensitive will get a bad headache from smelling sulfur dioxide, indicating an excessive metabolic influence on the head.

Sulfur dioxide can be found naturally in the soil. In a well-aerated soil, the oxidation of sulfur oxide will only produce such sulfur dioxide as can be taken up by plants in the form of sulfate; oxygen acting as the carrier of sulfur brings it into the cycle of life.

Sulfur dioxide is an antioxidant and an acid which is used to stabilise wine. It also has antiseptic properties and this makes it possible to select the micro-organisms before fermentation begins and thereby hinder the growth of undesirable microbes. Bacteria react to sulfur dioxide more quickly than to yeast.

As an antioxidant it combines with dissolved oxygen, hindering wine madeirisation and protecting it from excessive oxidation, whilst simultaneously retaining the desired aspects of flavour and aroma. The quantity of sulfur dioxide varies according to wine type and harvesting period. The total sulfur dioxide content of organic wine is less than that of conventional wine and in biodynamic wine it is even less. There are also wines produced today without the addition of any sulfur dioxide. Wine yeasts naturally produce small amounts of sulfite.

If it is necessary to add sulfur then its origin also plays a role. Most sulfur is currently produced from sedimentary sulfur. A large proportion of this sulfur is produced industrially from fossil fuels such as oil and refined gas. Sulfur extracted from volcanic rocks is sulfur that comes directly from nature, although it usually has to be purified before it is used. The sulfur used medicinally is of volcanic origin. This applies particularly to the anthroposophic medicines. The quality of a substance always depends on the way it has been produced.

13

THE NATURE OF COPPER

Marc Follmer

Metals occupy a special position among the natural substances. On the one hand, there are naturally occurring substances like potassium, sodium, calcium and magnesium that tend towards the Salt process, and on the other, substances like silica, phosphorous, carbon and sulfur that tend towards the Sulphur process (see previous chapter). In between there are the Mercurial elements that we refer to as metals (see Figure 70 and Table 3 overleaf).

Traces of the metals are found everywhere. Besides ore, iron and copper are also found in the soil and in living organisms, an indication of their biological value.

Gold, silver, copper and iron are found all over the earth, but only in a very few, specific places are they found in large amounts. Each metal has its own particular distribution across the earth. The metals have long been held in a group of seven, and the metals lead, tin, iron, gold, copper, mercury and silver have been connected with the seven classical planets in many traditions.

The metals have certain properties in common, such as electrical and thermal conductivity, ductility, light reflection and metallic shine. They are also heavy; their density is more than double that of water and in some cases even seven times more. In this way they represent a unique mixture of special qualities which are simultaneously terrestrial and 'super-earthly' (for example, in their connection to light and sound). These properties have been a source of fascination for humankind since time immemorial. Human development has been accompanied by the use of metals. We refer to the Copper Age (Chalcolithic), the Bronze Age or the Iron Age. Each metal possesses its own unique capacities and properties.

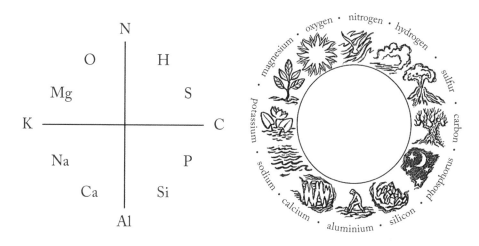

N nitrogen: the predominant atmosphere gas

H hydrogen: gas related to the extra-earthly realm

S sulfur: companion of vulcanism

C carbon: bearer of living substance

P phosphorus: bearer of the image of the stars

Si silicon: crystalline creator of stones

Al aluminium: creator of stones, cultivated soil

Ca calcium: sea salt, limestone

Na sodium: sea salt, blood;

K potassium: sea salt, plant sap

Mg magnesium: bearer of greenness in leaves, sea salt

O oxygen: air activated by the sun

The physical properties of copper

Copper is a chemical element belonging to the same family of metals as silver and gold. Like the other seven key metals, copper occurs naturally in the earth's crust and is important for the development of life.

Copper and its many ores appear in very intense colours. No other metal produces so many different colours. The pure metal has an orange-reddish tone, the colour of its ore ranges from gold-yellow (chalcopyrite) and blue (azurite), through to the green of malachite. With its many colours, copper

Figure 70. Selected chemical elements in their earthly aspects.

reveals its affinity to light. Its high thermal conductivity demonstrates its affinity to warmth. It is the best conductor of warmth after silver and also has extremely good electrical conductivity.

Thanks to its soft and pliable consistency, copper is easy to work. Its inner plasticity and malleability are the qualities of a precious metal, whilst its tensile strength and stability brings it closer to iron.

Copper is one of the metals that humankind has used for a long time. Archaeological evidence has shown that copper was being used before 5000 BC. The history of copper use in the Mediterranean region centres on the island of Cyprus where early copper mines were established. In the Roman period copper was called *cyprium* – metal from the island of Cyprus.

Metal	Silver	Copper	Gold	Iron	Tin	Lead	Mercury
Electrical conductivity	100	93	67	16	14	7.8	1.5
Thermal conductivity	100	91	74	13	15	8	2
Tensile strength	18	22	10	25	2	9	-

Table 3. Relative properties of the seven metals.

Chemical properties of copper

Copper appears in numerous chemical combinations and some remarkable alloys are possible – with tin it becomes bronze and with zinc brass. This latter was already known to the Greeks and bronze brought about a technological revolution (Bronze Age) around 2300 BC.

The wide range of possible chemical compounds shows a sensitive reactive capacity. This led it to its application – as copper sulfate solution – in the copper crystallisation technique for qualitative testing of plant juices.

Since ancient times, copper, with its various properties, its relationship to colour, its shine, its warmth, its malleability, has been connected with the Greek goddess Aphrodite, known by the Romans as the goddess Venus.

The island of Cyprus was dedicated to Aphrodite; Botticelli's painting, *The Birth of Venus*, perfectly shows the many different aspects of copper.

Copper and the living world

Copper is an essential trace element for micro-organisms, plants, animals and human beings.

The human body normally contains 1.4 to 2.1 mg of copper per kg (1¼ oz to 1¾ oz per 50 lbs), and is present as a trace element in many enzymes.

Plants need copper for photosynthesis. A copper deficiency results in the breakdown of chlorophyll, leaf die back and a reduction in growth. However, all lower plants, single-celled creatures, lower fungi and algae, are poisoned by copper.

Molluscs and mussels breathe with the help of pigments which don't contain iron but copper (hemocyanin instead of haemoglobin).

Figure 71. Sandro Botticelli, The Birth of Venus, *c. 1485, Uffizi Gallery, Florence.*

These organisms live in moist surroundings, are of a soft consistency and reject everything of a mineral, structure-giving quality.

Finally, copper has an anti-bacterial and fungicidal property. The copper alloys brass and bronze also possess such sterilising properties.

The copper process

Already with these few examples the gesture and subtle role played by copper in the processes of nature and the human being (and there are many others) becomes visible:[1]

- Copper is the metal of relaxation and warmth.
- In the human organism, copper works as a catalyst in metabolic growth processes.
- In looking at the human being we can connect the various character types with the different metals. The copper type expresses itself in humility, altruism and the capacity to adapt.
- In anthroposophic medicine, copper is prescribed as a treatment for cramps in combination with plants having the same properties, such as chamomile, lemon balm and tobacco.
- The kidneys and thyroid glands are the organs towards which the copper activity in the organism is directed.
- Its colourfulness and changeability and its ability to create alloys are characteristic properties of the metal copper.
- Venus / Aphrodite as the goddess of spring, beauty and love, bestows joyfulness and charm. While iron accompanies the brave, mercury the quick-witted and tin the good, copper belongs to the beautiful.

Copper in viticulture

Copper salts – especially copper sulfate, but also certain oxides, hydroxides and chloroxides together with sulfur – are some of the very few fungicides permitted for use in organic and biodynamic agriculture.

There are very few alternatives available. The most common form is the Bordeaux Mixture made up of copper sulfate neutralised with lime. The slaked lime neutralises the acid content of the copper sulfate solution. Copper sulfate also provides a supply of trace elements in situations where copper is deficient in plants and animals.

Like all metals, copper is broken down biologically in the soil, otherwise it could not be 'naturally' present in the soil. The accumulation of copper in the top soil layers can become critical.

The European legal limits have been changed over the years as follows:

- 🍇 8 kg/ha (7 lbs per acre) until 2005
- 🍇 6 kg/ha (5¼ lbs per acre) until 2014
- 🍇 4 kg/ha (3½ lbs per acre) after 2014

The aim must be to reduce the amount of copper used and to search for and find alternatives. Changes to the legal framework would then also be needed so that the alternatives can be applied. It is also necessary to breed plants with an in-built resistance. The application of plant teas (stinging nettle and horsetail, for example) makes it possible to reduce the amount of copper quite significantly (see Chapter 11).

The toxic effect of copper can be reduced by applying the total amount in several applications. A total application of 4 kg/ha (3½ lbs per acre) of copper, for example, can be given in 8 separate applications and 0.5 kg/ha (½ lb per acre) will not endanger the earthworm population to the same extent. Research is currently under way to assess the risk to the soil biomass of copper poisoning and determine how far copper applications can be reduced without compromising their fungicidal effect on mildew. Early results indicate that successful treatment depends less on the quantity of copper applied and more on its focused and repeated application.[2]

14

PRUNING TO REFLECT THE VINE'S NATURE

Hans-Christian Zehnter

A system that focuses primarily on yield and the mechanisation and rationalisation of work, has led viticulture into a one-sided monoculture. As a result, there has been a growing alienation between wine growers and their plants. This is reflected in the approach to pruning and the need to be as quick and efficient as possible. This has led to more carelessness instead of respect and attention to detail. The vines are weakened by this lack of care in pruning and are rendered more susceptible to disease, especially diseases afflicting wood.

A more gentle, or respectful, approach to pruning, would give consideration to the nature and requirements of the plant itself and not focus primarily on the economics of the system. The care taken would then ultimately benefit the business, not only in an economic sense but socially as well. Pruning in this way demands more exact and sympathetic observation – one can and indeed should try, for example, to imagine how the sap flows through the vine. Rudolf Steiner recommended developing a living understanding of the way formative forces are at work in nature:

> One can particularly help oneself in pursuit of this goal by
> observing the life of nature with inner heart's involvement.

One seeks, for example, to look at a plant in such a way that
one not only takes up its form into one's thoughts, but also, as it
were, feels along with its inner life, which stretches upwards in
the stem, spreads out in the leaves, opens what is inside to what
is outside with its blossoms, and so on. In such thinking the will
is also present in gentle resonance; and there, will is a will that is
developed in devotion and that guides the soul; a will that does
not originate from the soul, but rather directs its activity upon the
soul. At first, one quite naturally believes that this will originates
in the soul. In experiencing the process itself, however, one
recognises that through this reversal of will, a spiritual element,
existing outside the soul, is grasped by the soul.[1]

Those who carry out the pruning come into a significant relationship
with their vines. They find that their work has purpose and makes more
sense (and not only with the goal of making money). Apparent side effects
such as these can ultimately lead to an improvement in the atmosphere
of the whole undertaking, the way it is experienced and the quality of
the wine too, right through to the way it tastes. These are the principle
experiences of the two main representatives of 'gentle pruning': François
Dal in France, who describes his approach in the following chapter, and
the team around Marco Simonit and Pierpaulo Sich in Italy.[2]

The basic principle of species-specific vine pruning

This primarily concerns winter pruning. The growth habit of the vine is
acrotonic – growth from the leading bud is favoured, which means that
if left unpruned the vine grows rapidly away from the vine stock. In
the interests of rationalising the work the vine is therefore continually
pruned back close to the vine stock. This pruning is done in winter. The
gentle pruning technique is dealt with in more detail in the next chapter,
but it is worth mentioning two important aspects here.

Firstly, no large wounds should be inflicted since this hinders the plant's
flow of sap, causing it to close the wound on the inside by creating a cone
of dead wood. The dead wood however is an invitation for fungi to break

down organic matter. Since cuts made in old wood do not heal so well, the gentle pruning procedure limits itself to the pruning of shoots that are one or two years old and therefore some distance from the vine stock.

Secondly, pruning should be carried out in such a way that all the lines of flow remain functional. A careless cut can result in unnecessary sap flow blockages and diversions. The plant juices are then forced to flow along ever fewer pathways and the abandoned paths form dead patches in the wood. All of this weakens the plant and, in the end, causes it to wither and die. With careful pruning the supply lines remain straight and continuous and the wood remains unscathed.

15

GENTLE PRUNING TO PREVENT WOOD DISEASE

François Dal

Wood diseases are among the most unsettling challenges of the wine grower. Dead wood is almost always found inside vine stocks that are more than ten years old, even though they may appear outwardly healthy. The symptoms and occurrence of disease vary – often quite considerably – from vineyard to vineyard or from vine grower to vine grower. It may of course be the result of a number of practices, but the quality of pruning seems to have a strong influence on how the symptoms develop.

The SICAVAC (Service Interprofessionnel de Conseil Agronomique, de Vinification et d'Analyses du Centre) in France sees two particular factors as favouring the spread of wood diseases:

- Insufficient reserves in the vine plant. A plant with insufficient reserves is less resistant and therefore less able to withstand damage such as wounds and infection.
- Disturbed flow of sap. Everything which causes the sap to flow more slowly makes the plant more susceptible to disease.

Pruning necrosis
Tissue dieback due to one-sided pruning

From a certain age the shoots growing on one side of the vine will be nourished almost exclusively by the sap vessels on that side. Such an imbalanced vine whose entire sap flows on one side is therefore poorly supplied on the other. If the vine is **damaged** on this weaker side (through pruners, insects or the penetration of micro-organisms), it will be unable to respond adequately, and ultimately dies off to varying degrees. It might even result in an important section of the vine stock becoming incapacitated.

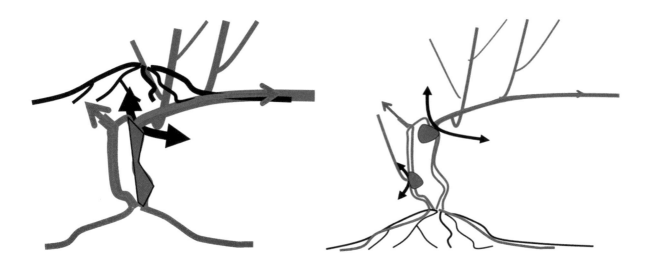

Tissue dieback due to pruning cuts

If a wound is caused by pruning, the vine responds by forming a cone of dead wood to prevent the plant from drying out. If the wound is so large and cuts into old wood, the cone will grow even bigger making it more difficult for the vine to prevent drying out.

The lack of balance and the various necrotic zones, knots and swellings, can in the end lead to the death of some of the wood, which of course results in a general weakening of the plant and creates conditions for fungal infection.

Figure 72 (left). One-sided pruning leads to a one-sided supply of nutrients.

Figure 73 (right). If pruning cuts are too large the vine is less able to prevent drying out.

Defence mechanisms of the vine
Compartmentalisation, CODIT model

When wood is damaged, whether through mechanical effects or undesirable micro-organisms, a complex defence response is set in motion. Tissue made up of lignin, suberin and phenol compounds is produced. This forms a protective zone which separates the damaged parts of the plant from the healthy ones. This phenomenon is known as 'compartmentalisation'. A model – the CODIT (Compartmentalisation of Decay In Trees) – was developed by Alex Shigo that sought to explain this mechanism. The formation of the tissue barrier is important in order to protect the damaged area from the surroundings. It has however two significant disadvantages:

- The plant requires a lot of energy in order to form the substances of the barrier.
- The tissue barrier irrevocably stops the flow of sap to a part of the trunk or branch.

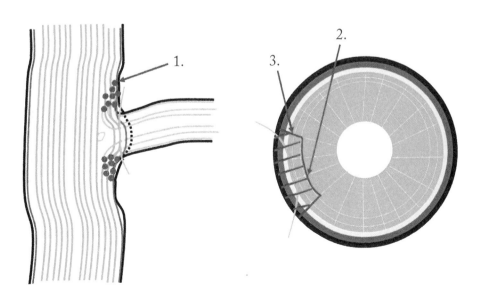

Figure 74. Left: Barrier 1 is formed by blocking the channels. Right: Barriers 2 and 3 are tissues in which starches are transformed into various substances like terpene, suberin, etc.

Cavitation and embolism

The formation of dead wood destroys surrounding plant vessels causing the flow of plant sap through the remaining vessels to correspondingly increase. When the pressure exceeds a certain limit, air bubbles are formed in the vessels. This phenomenon is known as cavitation. The air bubbles can block the flow of sap and cause an embolism. In order to prevent this the plant increases the viscosity of its sap by converting starch into sugar and thereby reducing the air bubbles. This process demands a lot of energy from the plant.

Conclusion

Injuries strongly curtail the flow of sap and cause the plant to draw on its energy reserves. This is why injuries lead to a general weakening of the vine and render it incapable of warding off pathogenic fungi.

Contrary to the widely held opinion that pathogenic fungi penetrate the vine and cause the wood to die, it is all too clear that the dieback of wood can be traced back to poor agricultural practices. The pathogens are secondary and appear in order to break down dead or weakened wood.

If the vines are weak and have insufficient reserves, they are also unable to produce barriers that are strong enough. Some particularly aggressive fungi are then able to break through these barriers and hasten the death of the vine stock.

Preventing necrosis with respectful pruning
Training young vines

Special care and attention need to be given to the young vine. Pruning injuries inflicted in the early years affect the development of the vine far stronger than hitherto believed. In order from the very beginning to develop a system of pruning which interferes as little as possible with the flow of sap, it is necessary to be continually aware of the growth and development of the vine. We must therefore introduce some new practices: when the young vine is trained to a wire, a longer shoot should be cut, bent and tied to it in such a way that the newly sprouting buds can continue growing along the wire.

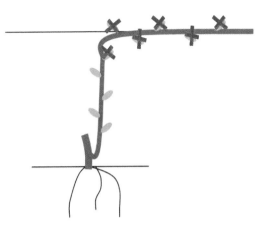

In the following year it is already possible for two short fruiting spurs to be left. **Both ar**e tied to the wire even when the shoots later used as fruiting or renewal spurs are not growing in the direction of the wire framework.

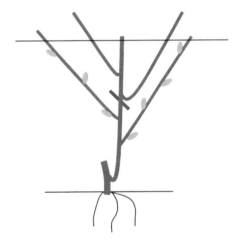

The aim from the very beginning is to allow the vine to grow in a natural way. This requires giving careful thought beforehand as to how this form **can be maintained**. And it should be clear from what has been said previously that extensive pruning injuries and one-sided developments are to be av**oided**.

Guyot Poussard's pruning method

Several approaches to pruning lead to very satisfactory results. For example, goblet training, double cordon and the Guyot Poussard method can be mentioned. With all these approaches however,

Figure 75 (top). Training shoots so that newly sprouting buds continue growing along the wire.

Figure 76. Two fruiting spurs in the following year.

we should never lose sight of the principles described above. It is particularly important never to prune too close – always leave a small piece of stem behind which can then dry out. By paying attention to these principles the flow of sap will not be hindered. It is always better to leave a somewhat longer stem or leave an extra bud or two in place than to prune too closely.

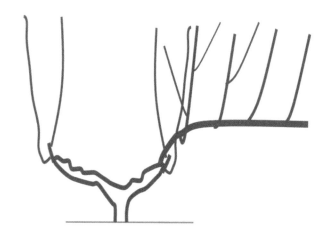

Other factors preventing accumulation of reserves or flow of sap

Well-known and tested factors which support the build up of reserves and the flow of sap are:

- 🍇 A well-balanced harvest yield (not too much).
- 🍇 A balanced vitality (neither too much nor too little).
- 🍇 A living soil that is in equilibrium. Existing nutrient deficiencies can be corrected and the vitality optimised through judicious manuring. A green soil covering has many advantages in terms of soil life and contributes, through the build-up of reserves, towards keeping the vine in good health. Green cover also helps to strengthen vitality so that it is present in the period between budburst and flowering.

Figure 77 (above). Guyot Poussard's pruning method.

The flow of sap can be disturbed by secateurs mutilation. There is however another factor which can seriously impinge on the flow of sap. This concerns the quality of the graft between rootstock and scion. Back in 1895 B. Drouhalt wrote: 'The vitality and longevity of a grafted vine depends largely on the quality of callus that is formed. A poorly healed grafting point may well produce strong vines in the first few years, but over time the vine will sicken and eventually keel over. All kinds of possible reasons are sought to explain these vine deaths but the problem is basically the poor graft union.'[1]

If every vineyard were laid out in such a way that the vine plants could build up maximum reserves and their sap could flow without being impeded, the mortality of vines caused by wood disease would significantly fall.

A respectful system of management such as is here described can also have a positive effect on other potential afflictions of the vine (symptoms of deficiency included). Furthermore, vines which are better nourished have more reserves, and vines whose sap flows in a healthy way will suffer less from adverse weather conditions and need less fertiliser.

It may be that all these measures require additional work. The increased work however will result in better and more balanced grapevines and will require less time to be invested in the treatment of disease.

16

THE VINE NURSERY: TRAINING THE BREEDER'S EYE

Jean-Michel Florin

The wine grower at Domaine Le Sang des Cailloux in Vaqueyras, France, said, 'I have completely changed my cultivation methods. We are planting the Rupestris du Lot rootstocks and then, in two or three years' time, they will be grafted on site using scions from our own mass selection. The vines are more resistant and the sap flows better.'

In viticulture, as in many other fields of agriculture (such as cereal production), the work of plant breeding was taken away from producers during the last century. The breeding of new vine varieties was taken over by specialist institutions. The grape louse (phylloxera) crisis at the end of the nineteenth century sped up this process because it suddenly became necessary to buy grafted American rootstocks. It is possible to ask oneself whether the phylloxera plague was in fact an excuse to take vine propagation out of the hands of wine growers. Lucien Daniel, for example, pointed out that solving the grape louse problem through the grafting of American rootstocks opened up a whole new market that included rootstock sellers and sugar producers.[1] The question should be asked as to why the only legally permitted solution is to graft on to American rootstocks.

There would surely also have been resistant European varieties to be discovered and bred from. However, **what is** clear is that the solution chosen removed from wine growers the possibility of breeding and propagating their own vines.

Previously, every wine grower **propagated their** own vines – mostly from cuttings taken on site. Although this method of vegetative propagation did not bring about diversity to the same extent as generative propagation via the seed, it nonetheless resulted over time in an adaptation to the site. In this way each region developed its own site-adapted plants of the various vine varieties.

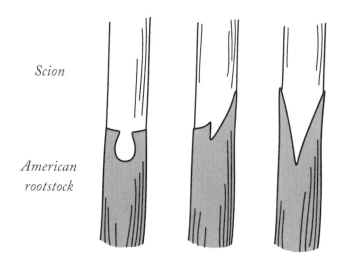

Scion

American rootstock

In most of the plant breeding institutes the approach to propagation and the grafting of European vines on to American rootstocks is very highly mechanised, primarily in order to reduce the costs. The root stocks are mostly grown conventionally and with a lot of artificial fertiliser and chemical sprays to speed up growth. They are often grown directly on bare soil without any supports in an environment that tends to be too damp and shady. This in turn makes the plants susceptible to diseases of decay like botrytis. Fungicidal sprays are then often used to control botrytis. Apart from this, the root stocks often suffer wood damage by being carelessly walked on or even driven over by a tractor, which in turn makes them more susceptible to disease. These root stocks are often quite weak too, the wood

Figure 78. Grafting methods left to right: omega method, whip and tongue method, cleft (V) method.

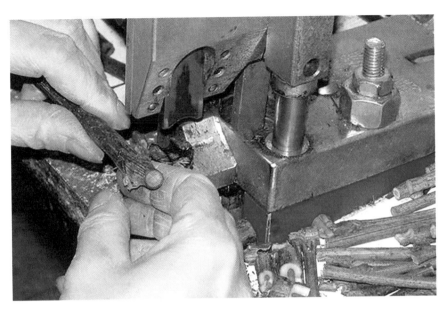

Figures 79–83. Stages in the modern process of vegetative vine propagation in the vine nurseries. Using an omega cut the rootless shoots of the American vine are joined to the scion of the European vine. The graft is sealed with paraffin. The grafted cuttings are then planted outside in vast numbers so that they can form roots and develop vegetative growth above ground. These mass-produced vine plants are then offered for sale.

is not very firm, contains a lot of **pith,** and is sensitive to wood disease. Due to the use of fungicides these plants are often not able to develop any mycorrhiza around their roots.

The root stocks from just a few American species are not adapted to the particular site. There are currently between thirty and forty clones across the whole world, which are being cloned again and again. The scions are likewise cloned – purely to achieve greatest possible uniformity.

The highly mechanised omega grafting procedure is often carelessly implemented because of the need to work fast. The omega method is three times faster than the whip and tongue method. (A study undertaken by Florian Bassini showed that plants with omega grafts were more susceptible to wood diseases; see Chapter 10.) After they have been grafted, the root stocks are dipped in paraffin, then in a rooting hormone, and afterwards again in paraffin before being planted out in the nursery or in pots.

All this has led a number of wine growers and propagation units to follow a different course.

Guy Bossard, a wine grower from Nantes in France, has been running a vine nursery for more than fifty years. During this time he has been breeding his own vines and since 1997 has used biodynamic methods. Having carried out a mass selection he only chooses scions from healthy vines with an average yield, loose bunches of grapes and resistance to flower drop. He cuts the scion from well matured canes during the ascending moon just before they shoot. The selection must be large enough to prevent a decline in diversity. The fact that breeding has focused solely on yield over the last forty years means that little has been done to improve quality, especially since selections were generally made from plants that were weakened by producing high yields. Guy Bossard grafted with the whip and tongue method and used beeswax instead of paraffin to seal the wounds. Instead of rooting hormone the grafted plants were soaked in biodynamic horn manure preparation. 'We graft our plants using scions that we have selected and apply the whip and tongue method of grafting. It is certainly not as quick as in industrial vine nurseries, but it is efficient and is carried out with a good conscience. The results are very good – very little failure. And we keep our independence,' said Guy.

A first step towards enabling wine growers to take control of vine breeding once more, would be to develop closer cooperation between wine growers and vine nurseries, which is already happening here and there. A further step is to bring the vine nurseries back into the vineyards as is done with biodynamic cereals; breeding is to a large extent carried out on biodynamic farms.

Figures 84 and 85. Wine growers like Lilian Bérillon in the south of France are following their own path. They apply biodynamic preparations, pursue mass selection to retain the genetic diversity of European scions and raise the American rootstocks themselves.

Grafting

The grafting procedure unites two different plants and subsequently grows them as one. It is no trivial act and it should be carried out conscientiously.

By far the most widely practised grafting method today is omega grafting. It is very easily mechanised and can therefore be done very quickly. It comes with problems however – it has the reputation of encouraging wood disease infection. The flow of sap between scion and rootstock is not very good, causing many necroses to form.

The continuing increase in cases of wood disease and the premature death of grape vines has given new impetus to the search for new grafting methods which can better facilitate the flow of sap in the way the whip and tongue or cleft (V) methods do. These methods were once used quite widely but have been forgotten because they take longer to carry out.

In the following contribution, François Dal describes in some detail the steps needed and the methods required to integrate the practice of grafting into one's own vineyard.

17

REGENERATIVE GRAFTING

François Dal

Vines affected by esca (see Chapter 10) are often found to have a healthy rootstock. In these cases, the vines can be maintained or regenerated through grafting.

There are a number of different grafting techniques. The two methods most commonly used on vines are chip budding and T-budding. In both procedures a dormant bud is inserted in the rootstock so that it is in direct contact with the cambium layer.

As mentioned, these two methods have both proven themselves to be effective, although they are difficult to apply on vines afflicted with wood disease. In any case, newly grafted vines demand special attention and care (regular watering, bud removal, etc.). The best solution appears to be the complete removal of the upper plant and to retain the rootstock. Regrafting using the cleft technique appears to be the most suitable way of regenerating existing plantations with wood disease.

The regrafting procedure using the cleft technique needs to be carried out when the sap is flowing strongly – but not too strongly. The ideal time would seem to be in spring just before budburst. In more northerly regions (of the northern hemisphere) satisfactory results have been achieved between mid-March and the beginning of June. It is also possible to carry out grafting in the autumn. If grafting is to be done shortly before blossom time, the sap should be allowed to drip for a few days beforehand. The right moment needs to be found: if grafting is carried out too soon there is a risk of drowning the entire graft in sap; if it is done too late the plant may already have lost its vital power.

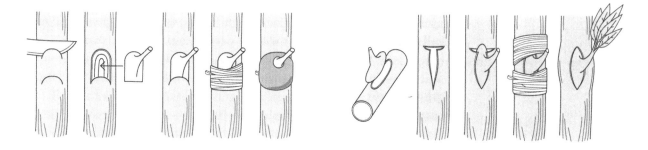

The harvest and conservation of scions

The harvest and conservation of scions is an extremely important procedure, but it is only necessary if regrafting is carried out after budburst. When regrafting is done at the end of March or the beginning of April, scions can be taken from vines that have not yet been pruned. The scions must be in perfect condition at the time of grafting and have not yet started sprouting. They therefore need to be harvested at the end of February or beginning of March, that is before there is an intense flow of sap. Harvest should take place on a day without rain to ensure the wood is dry. The buds must not be allowed to dry out, go mouldy or start sprouting.

The slips for grafting are tied in bundles, packed in polythene bags with plenty of breathing holes and stored in a cool (3–5°C, 37–41°F) place (a cellar or fridge). The twigs should not be too thick (6–8 mm, ¼–⁵⁄₁₆ in, in diameter), ideally without many **tendr**ils and no side shoots. Scions are best taken from young vine plants (4–6 years old) in order to reduce the possibility of viral infections. The condition of the slips should be checked at least every two weeks.

One or two days before grafting, **the bund**les should be put in a bucket of water (like flowers in a vase) in order to re-moisturise them. The room temperature should be between 12 and 15°C (54–59°F). The quality of the

Figure 86 (left). The Chip budding classical grafting cut.

Figure 87 (right). The T-budding classical grafting cut.

wood and buds should be tested by taking a few of the twigs and cutting into them – the cuts should be of a beautiful green colour. If, however, the wood is a faded green or grey, it means it's too dry and no longer usable. If the buds are brown, they are of a poor quality and must be discarded. After 24 hours in water the buds should start to swell. Slips whose buds do not swell, should not be used.

Grafting technique

First step: cutting the crown

The earth around the rootstock must be removed. The vine must be cut below the existing graft and at least 5 cm (2 in) above a node on the rootstock. A tool (such as a hand saw) should be used that enables a clean cut to be made. If it is not a clean cut it should be tidied up and made smooth with a knife, especially at the place where the new bud is to be implanted. Should the graft site not be entirely healthy, a cut can be made lower down. There is no problem making a graft 3–5 cm (1–2 in) beneath the earth.

The cut can be made several weeks before grafting, but in that case the cut will **need** refreshing again before the graft is made.

Second step: making a cleft in the rootstock

With the help of a thin but strong tool, the rootstock is split. A knife and a hammer can be used for this purpose. The cleft should be deep enough (3–4 cm, 1½ in) to insert the scion. It should be made at the centre of the rootstock where the diameter is at its largest. Ideally, the wood in the cleft should not splinter.

Third step: preparation of the scion

The preparation of the scion is the most critical stage and the source of many mistakes, especially if the diagonal cut is not carried out correctly. It is extremely important to have a clean cut for the graft. The scion should be cut to a sharp point. The cuts on both sides should start underneath an eye. Each cut should be very precise and carried out in one go. If the cut needs correcting it can easily happen that the surface is not smooth and this can lead to a poor contact with the rootstock.

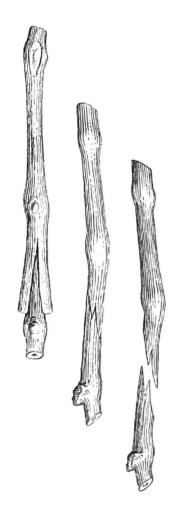

Figure 88. Recommended grafting cuts for the vine. Left: cleft; centre and right: whip and tongue.

The two sides of the wedge should be straight and not bent. To achieve this the scion must be held in such a way that when cutting, your wrist does not need to turn, only your elbow moves. After the first cut away from you, raise your arm and hold your thumb along the first cut. The second cut is done upwards on the opposite side, and your thumb can follow the cutting movement. This is a way of ensuring that the wood is cut correctly.

The scion need not necessarily end with a leading bud. The thickness of the remaining wood must be greater on the side with the eye than on the opposite side, so that the best possible contact can be achieved between rootstock and eye. The scion should have a length of more than 10 cm (4 in) so that there is space for two or three buds.

The scions can be prepared in the morning for using all day. They should in this case be kept in a bucket and covered with a damp cloth to prevent them drying out.

There are of course machines which can very quickly create the scions, they ensure that wedges have the correct angle. Their knives however are very difficult to sharpen – and of course nothing can be better than good work carried out by hand.

Fourth step: inserting the scion into the rootstock

It is very important that the cambium of the scion is brought into close contact with the cambium of the rootstock. The larger the area of contact, the greater the chance of a successful graft. Because the cambium lies immediately beneath the bark the scion and rootstock must fit tightly bark to bark. In an ideal case the scion and rootstock are so seamlessly joined that only by running your hand over them is it possible to discover where they were brought together. The grafting of a second scion – one on each side of the rootstock – is strongly recommended since it increases the success rate.

The first eye on each scion should be on the outside of the rootstock. The buds produce hormones which promote the union of the two cambiums. The first eye is found close to the grafting point. A suggestion for making the insertion of the scion easier is to hold the cut open with a tool (such as a screwdriver).

Once the scion has been inserted, it is important to check the quality of contact between the scion and rootstock. This means bending down in order to have a really close look at the join. If the angle isn't correct or the scion not properly positioned, the chances of success will be very low.

Fifth step: protecting the scion

The most effective way of protecting the graft is to have a protective cover filled with soil or sand. This protection has three important functions:

- 🌰 To prevent it drying out. The graft is very sensitive to drying out. The sand or soil keeps the scion moist.
- 🌰 To reduce the temperature around the graft and make it more stable. This slows down the sprouting process so that a full knitting-together occurs before the buds burst.
- 🌰 To protect from mechanical injuries.

Ideally there should be a plant stake (or two if the ground near the rootstock is being cultivated) and a protective cover at least 10 cm (4 in) in diameter. It is a useful protection against rabbits and herbicides as well as helping to encourage an upright growing habit. The protective cover is filled up to the second bud (or 10 cm / 4 in), with soil or a material that holds moisture, such as sand, wood shavings or vermiculite. If the soil used for filling contains a lot of clay, it should not cover the second bud. Alternatively, the covering can be 1 or 2 cm (½–¾ in) high. The soil in the protective cover should then be lightly pressed together.

The incision in the rootstock can be covered to prevent soil from entering. A wound-sealing ointment (which also has an anti-drying effect) can be used to do this, or alternatively a couple of sheets of absorbent paper can be laid over the cut before the protective cover is filled with soil.

If the graft develops correctly two or three shoots will develop and even grow to 1½ to 2 m (5–7 ft) in the first year.

Advantages and disadvantages of grafting

Thanks to this technique, vines affected by esca are, after one year, able to produce half of their normal yields. Normal levels of production are then reached after two years. As a comparison, newly planted vines will produce half the expected yields at the earliest after four or five years and rarely reach full production before six years. The regrafted vines retain the rootstock and hence the age of the original vine.

This gives a net gain in quality because ultimately the time needed for grafting is less than the time and effort required for planting new vines, quite apart from the higher costs involved in their purchase.

Depending on the precision with which the work is carried out and the weather conditions in spring the success rate can lie between 80 and 90 per cent.

The biggest disadvantage in this technique is that the most favourable time is spring – the very time which should be occupied in training, bud removal, etc. Grafting in this period therefore requires work to be well organised. Another smaller disadvantage is that although this procedure is simple it requires great care in carrying it out and not everyone is able to do it and achieve the same degree of success.

Care of regrafted vines

The vines whose grafts have developed well can grow 1½ to 2 m (5 to 6½ ft) in the first year. It is best to tie in these canes. The graft normally starts forming roots during the first year. These should be removed together with the protective covers when the new shoots are 40 to 50 cm long (16–20 in) around early July. It is advisable however to put the protective cover back on, but this time without soil. If it is not possible to do this work in July, the roots can also be removed in winter.

When pruning, it is best to keep only one of the scions which have grown. It can otherwise happen that the two scions interfere with one another as they grow. Depending on its vigour the scion should be cut back to between four and eight eyes. There may also be more eyes since the scion already has the root system of a fully-grown vine.

18

GROWING POINTS

Jean-Michel Florin

As we saw earlier the growing point of the vine, or the apex of each shoot, is a very important organ (see Chapter 2). It is a regulator and, at the same time, a kind of sense organ that enables the vine to connect to its surroundings and also protect itself from external influences. These growing points are usually trimmed back in summer encouraging the growth of axle buds.

Daniel Thulièvre, a biologist and winemaker who has worked with biodynamics for fifteen years, explained during a biodynamic viticulture conference why trimming the growing point of the canes is problematic, and described the alternatives he has explored.

If the shoot tips are cut, the vine is weakened and left for a few days without any means of resisting pathogens. The vine is in fact traumatised.

From the bursting of the buds to the moment of flowering the vine's growth is predetermined. The shoots grow by remembering, as it were, the previous year. Only after eight nodes are there new organs. This is when the growing point opens itself up to the current situation, which includes climate, pests and diseases. Each shoot is in some way autonomous and the apex or growing point makes its own response to the current conditions of growth. If this apex is cut, the shoot loses the possibility of absorbing information about the year's cycle. The growth of the roots is also intimately connected with the shoot growth. If the tip is removed, root growth is also impaired.

The trimming of shoot tips can be replaced in various ways. One way is to raise the height of the espalier frames. The canes can also be 'bundled'. This involves taking two groups of five canes and making a simple knot with them. The canes are then rolled together along the top wire of the frame. After two hours the tendrils will have done their work and the knots are firm. This procedure is only carried out once during the year – in western France (Anjou) this takes place in mid-July.

The results seem satisfactory: there are less diseases and pests because the rows are less dense and the vines receive more air and light.

PART 5

The Future of the Vine

19

WINE GOLD

Hans-Christian Zehnter

In this chapter the aim is to look at wine growing perspectives that, beginning with the healing properties of the vine, go beyond the predominant use of the vine as a means to prepare wine as an alcoholic beverage.

It has been repeatedly stated in earlier chapters that wine is far more than the product of the vine. The term *terroir* has been mentioned in this connection several times. It refers to the character, uniqueness and value of a locality, and relates these factors to agricultural products, especially wine. The *terroir* of a wine can therefore be seen as a self-contained, individual organism that infuses the whole process of wine growing and wine making, from the cultivation of the vines to filling the bottles with the wine. In relation to Goethe's method of observation of nature, Rudolf Steiner wrote:

> It must be admitted that all the sense-perceptible factors of a living being do not manifest as a result of other sense-perceptible factors ... All sense-perceptible qualities manifest as the result of a factor that is no longer sense-perceptible. They manifest as the result of a higher unity hovering over the sense-perceptible processes.[1]

The painting, *Village Landscape with a Flowering Tree* by Cuno Amiet shows in a wonderful way that the flowering tree radiates out over the whole village – over the houses, the corn field, the sky and even the time of year. The flowering tree is actually the whole situation, the entire context. It is released to a certain degree from the object 'tree' into the mood or atmosphere of its surrounding landscape – or if you like into its *terroir*.

A linden or lime tree is very different in summer to what it is in winter. In wintertime it is surrounded above all by expectation; by the expectation that in summer when it is flowering it will once again be able to radiate out into the whole sensory-terrestrial surroundings. In winter it has to a great extent withdrawn from these sensory surroundings and has transferred over into a soul-spiritual condition.

Let us now translate this into the situation of the vineyard. In winter it lies leafless before us, the gnarled grey vine stocks appear as though dead; only our memory gives us hope for another growing, flowering and fruiting period. During the spring this hope transforms itself into expectation. In the surrounding area nature is sprouting, turning green, making the atmosphere around the still-barren vines even more dense until even here, with the first shoot, life begins to stir and become visible. What has up until now lain hidden in the deep recesses of memory, increasingly spreads itself out across this place, finding its way in order then to fill the vineyard with an abundance of trailing vines.

What the vine produces as a fruit in the autumn is harvested by human beings, enclosed inside barrels, and there starts to ferment. It is precisely during this 'imprisonment' that the wine develops its site-specific characteristics. What was present as *genius loci* around the vineyard is now made available through human activity as taste and smell. What previously was a watery juice now becomes a 'spirited' drink. The winemaker is thus capturing the supersensible nature of the wine and making it available as a tincture; what had been atmosphere and mood is now substance.

These introductory thoughts have been brought in order to introduce the medicinal effects of the grapevine and of wine. It is in this context too, that much is revealed about the golden nature of wine -- and perhaps also of its future.

The vine is used in the most varied of medicinal preparations. For example, Weleda, an anthroposophic pharmacy known for its medicinal and cosmetic products, uses grape vine and wine in over thirty remedies. For instance, Kali Aceticum (potassium acetate) is a substance of central importance. During the course of a very specific production procedure, a kind of model is created for an ordered and well-mannered interplay of the four human members (physical body, etheric body, astral body, and 'I'). Kali Aceticum is used in situations where these relationships have been thrown out of balance, for example in adolescent psychiatry.

One of the most well-known wine-based medicines is Hepatodoron. It is made from dried strawberry and vine leaves and is primarily used for strengthening the activity of the liver.

Figure 89. Cuno Amiet, Village Landscape with a Flowering Tree, *1905, Kunstmuseum Solothurn, Switzerland.*

To do true justice to this 'wine gold' both in terms of its cultivation and breeding, as well as in its marketing, we need to develop new concepts to acknowledge its inner nature. The following chapter by Nikolaus Bolliger will offer insights in this direction.

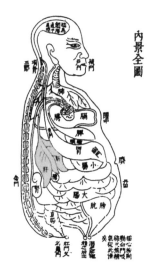

Figures 90 and 91. The strawberry (top right) and vine leaf (top left) form the basic components of Hepatodoron

Figure 92 (bottom). In traditional Chinese medicine, the liver is often depicted as a green leaf.

20

DEVELOPMENTS IN APPLE BREEDING: A POSSIBILITY FOR THE VINE

Nikolaus Bolliger

Apples are, by and large, cross pollinators. Only rarely does the offspring of an **apple** produce the same variety of apple. The only way to maintain and produce the special qualities of a particular apple variety over many generations is through vegetative propagation, namely by grafting. In doing so, however, the plant misses out on that vitally important moment of chaos in the seed (see Chapter 1).

This vegetative propagation and breeding leads to a reduction in diversity. Golden Delicious is the source variety for most modern breeding. It is now virtually forgotten that this variety of apple was a chance discovery, grown from a seedling in the 1890s when there was far greater diversity. Well known varieties such as Granny Smith and Braeburn are also chance seedlings.

It is clear therefore that when a plant is only propagated vegetatively for many generations it loses two possibilities for regenerating itself – variability and the passage through the seed stage. If we want to cultivate plants that remain vigorous in the long term, an approach to breeding is needed which takes these two sources of regeneration into account. Only in this way is it possible to develop new varieties which on the one hand satisfy today's demand for high quality fruit, and on the other, thanks to their health and vitality, require but little in terms of direct plant protection measures.

It is precisely these findings that are transferable to viticulture. Since ancient times the vine has also been maintained through vegetative propagation (see Chapter 4). There were only a few exceptions in which grape seeds were sown (for example by some monks in Burgundy) in order to select the best plants and then propagate them further from cuttings.[1] Following the grape louse plague of the nineteenth century, diversity was reduced even further to only a few clones. Recent exceptions to these are some of the PIWI, or fungus resistant, varieties.[2]

The new perspectives being developed in apple culture could therefore also be of interest to wine growers.

Classical breeding practices

It has been calculated that around 400,000 species of plants are growing on the earth.[3] In the course of time comparatively few of these species have been cultivated by humankind. The development of cultivated plants is so closely connected with human beings that without ongoing care their continued existence cannot be guaranteed. Cultivated plants would disappear if we no longer grew them. Along with cereals as the most important staple foods, grapevines count among the oldest of cultivated plants in Eurasia. They have been closely connected with humankind for ten thousand years. They spread gradually to all regions that were climatically suitable and developed over time into many local varieties.

The cultivation of wild plants led in the course of millennia of human and plant evolution to the development of cultivated varieties. From the connection developed between human beings and apple trees, thousands of local native varieties came into existence. It is often referred to as the time of *unconscious breeding*. I would prefer to call it *breeding with heart and hand* because loving care was the key. Cultivated plants and the food derived from them serve not only to stave off physical hunger but also form part of an all-encompassing totality which can be perceived and experienced in the rituals of thanksgiving practised by people of all cultures. This has also continued with the Christian ritual of the Eucharist with its symbols of bread and wine.

It was only about 150 to 200 years ago that the conscious breeding of apples began. Apple pips are sown and from the progeny, plants of the desired characteristics are selected. If the characteristics of two varieties are to be combined, the free pollination of the flowers has to be prevented. Pollen from the 'father' flower is transferred with a brush to the 'mother' flower. After it is fertilised it takes many years before the young tree is large enough to produce its own fruit. It then takes several more years for the new variety to be selected and tested. Once the new variety has been confirmed it is given a name and then propagated further in a vegetative way by grafting.

Genetic procedure

It was typical of the nineteenth century for every field of life to be investigated and repeatedly analysed. Scientific knowledge began to flow into the field of agriculture and brought change to traditional practices. Plant breeding lost its 'unconscious' characteristic. Scientifically trained plant breeders had clear breeding goals as they began crossing and making selections. The intellect came ever more strongly to dominate the breeding work.

This development led to the molecular-genetic approach to plant breeding that we have today. The super uniform plants that grow in the cultural deserts of industrial agriculture are, in terms of yield, far more productive than the old varieties, and yet they mark an end point in development. Cultivated plants whose fertility is prevented by terminator genes are not some horror scene out of a science fiction novel but the real consequence of the loveless relationship between *Homo economicus* and plants.

All over the world the use of GM technology (also called biotechnology) is promoted as a more efficient and less expensive way of meeting breeding objectives than classical methods.

Foreign genes, for instance, are being inserted into apple seedlings so that they can flower and bear fruit within a few months of being sown.

This dramatically speeds up the sequence of generations. Later on, the active gene can be removed again.

Many advocates of biotechnology maintain that GM breeding techniques, which only make use of genes from the same species, are unproblematic and should therefore be permitted. Organic plant breeding, however, completely prohibits all methods of breeding which change the genetic make-up in any other way than through the natural process of flowering and seed formation. This is quite independent of whether it involves foreign genes or the species own genes.

Separate consideration is needed for marker-assisted selection (MAS) in which DNA analysis is used to determine the presence of a certain gene so that subsequently a direct selection of plants with that gene can be made. This method is purely diagnostic, no genetic changes are brought about, and in certain situations it can be helpful and even save costs.

Biodynamic apple breeding

In this context we need to ask ourselves the question, how should our relationship with cultivated plants be developed so that the co-evolutionary process has a future? Is it possible today for the breeder to form a connection with plants that is both personal and evolving?[4]

Biodynamic apple breeding builds on classical breeding methods. It takes the view that the gene is not the origin of life but provides the material foundation for the processes of life. The stream of life which flows from the past into the future, leaves behind traces of its activity to be passed on as inheritance. The processes which take place in the flower after fertilisation and conclude with the formation and germination of seeds is, to some extent, a process of chaos in which the inherited material structures are dissolved before reordering themselves in a new way. This process gives the plant a new orientation out of which a whole range of new possibilities can emerge for its future development. What happens here is not only affected by the immediate surroundings but also by the laws of the cosmos. In order to do justice to all these processes it is important that plant breeding takes place in a well-managed and natural context. That is why a biodynamic farm is the ideal place for engaging in the further development of our cultivated plants.

Just as each day is informed by the rhythm of day and night, there are periods of growing and fruiting in the life cycle of a plant followed by a time when it rests in the seed. When the seed germinates a new cycle begins which culminates in the formation of new seed. During its period of growth, the plant expands out into the space around it by using its living activity to take hold of dead substances, giving them life and thereby manifesting its own form. Various salt substances in the soil are dissolved in water and carried upwards on the stream of plant sap. Using light and warmth the plant absorbs the invisible carbon dioxide of the air and carries it down to the roots in the descending stream of plant sap. Once it has flowered, the plant begins to concentrate its strength in the seeds as they ripen. For many plants this process is so intense that they can only form seed once, after which they die. Following this extrovert phase of growth, a shorter or longer period follows in which the seed lies dormant with no visible signs of life – until germination inaugurates a new cycle.

In the agriculture course Rudolf Steiner describes how these externally visible processes of life are connected with spiritual beings and the forces of the whole cosmos. The forces of the superior planets – Saturn, Jupiter and Mars – are reflected up into the plant from below and work via the silica-based substances in the soil, and from the periphery they work directly to support sugar formation, flavour and aroma. On the other side, forces from the inferior planets – Venus, Mercury and Moon – are drawn in by the calcium substances in the soil. They ray down through the plants from above and are met by the rising stream of dissolved salts. The forces of growth and maturation can be understood as an interplay between the forces of the superior and inferior planets.[5] These can be guided, ameliorated and strengthened with the help of horn manure preparation (growth, inferior planets, mediated by calcium) and horn silica (ripening, process, superior planets, silica).

In the seed state plant life is at a standstill, outwardly nothing is happening. The accepted position is that life is being stored in the protein molecules. Rudolf Steiner broadens this approach by adding that although this genetic material ensures that the same species grows out of the seed, it dissolves as it were into a state of chaos and, influenced by forces from the whole universe, creates itself anew: 'The new organism is formed out of the whole cosmos each time this seed-chaos state occurs.'[6]

These considerations indicate how important the process of seed formation is for the plant because each time there is a new cosmic orientation. With cultivated plants that are subject to vegetative propagation and therefore do not go through the seed process, the question arises whether they should perhaps not go through the state of seed chaos from time to time in order to renew and reorientate themselves. Do they perhaps lose **their** connection to cosmic forces over time and therefore also their vitality and resilience?

The unique nature of the vine

In few other cultivated plants is the influence of variety so decisive in terms of the characteristics of the processed product as it is with the vine. It is simply unthinkable to consider using any other variety than that which has always been used to produce a well-known traditional wine. This demonstrates the need to continue honouring the classical varieties. It brings up the question as to whether it is possible to revitalise and bring about a new orientation by using measures other than seed formation.

It does, however, also make sense to develop new adapted varieties via the seeding route, and to use cross-breeding techniques. If there are significant changes to the environment – whether because a variety is being grown in a region with hard winters or because of introduced diseases – the most long-lasting way of solving the **problem is through** plant breeding. That the typical characteristics of vegetatively propagated cultivated plants like the vine will thereby get lost, is unfortunately unavoidable. The breeder can, however, select from the diversity among the offspring and choose improved individual strains that **are more** vigorous, healthier and better adapted. This is how the new arises, and new vine varieties can also open up new and interesting wine flavours.

21

THE FUTURE OF WINE: WORKING WITH A CREATIVE TENSION

Jean-Michel Florin

There are many contradictory demands placed on biodynamic viticulture and those engaged with it have to find their own independent paths:

- Between an industrialised and globally intense monoculture (that is also furthest away from the idea of the farm organism) and a plant which is predestined to express the individuality of the site upon which it grows.
- Between a special branch of biodynamic agriculture which contributes greatly to public awareness and acceptance of this approach, and a product which is actually more of a luxury than a food.
- Between a cultivation method that works in harmony with nature, cosmos and the human being and the danger of biodynamic increasingly becoming a one-sided elite brand.
- Between a plant which on the one hand is very responsive to the biodynamic spray preparations, but whose vitality on the other hand is greatly weakened.

🍇 Between a 'tasting' culture extending from *terroir* through to a highly refined capacity for living awareness and expression, which sets an example for other fields of agriculture and highlights their advanced lack of knowledge about the biology and nature of their own cash cow, the vine.

Seen in this way the vine encourages or even demands an individual quality both in the grower and in the landscape. Perhaps this is connected with what Rudolf Steiner said about the effect of the vine upon the human self, the I?

The atmosphere created by the vine also helps to balance out bad effects. As you know, lime blossom is quite powerful, and walnut trees also have a powerful scent; this has more of a vitalising effect on the [soul]. And the atmosphere created by the vine has more of a vitalising effect on the I. So there you get a powerful effect also on the higher [parts of the human being].[1]

It is important to note that Rudolf Steiner is speaking here about the vine and not about the alcoholic wine drink. Alcohol in particular hinders self-development. It helped people in earlier, pre-Christian times to become truly at home on the earth. For the spiritual development which is called for today, alcohol is certainly not so helpful. This does not mean that the grapevine will not have a task in future. It already has many: as food or, for example, the many medicines made from different parts of the plant – leaves, grapes, juice, pips and the wine itself. The vine is a plant with a great deal of vitality – and because of this, it will conceivably become even more significant. It has huge potential. Discovering this future potential should become one of the main concerns of biodynamic viticulture and biodynamic research.

Figure 93. Wine harvest at the Champagne Piollot, France.

ACKNOWLEDGEMENTS

The idea for this book arose out of the first international wine-growing conference that took place in Colmar, France and in Dornach, Switzerland in 2012. A larger project developed out of this initial impulse which resulted in further French-language conferences organised by the French biodynamic movement (Mouvement de l'Agriculture Bio-Dynamique) between 2013 and 2016. We would therefore like to thank all the authors and contributors to these wine-growing conferences who have given us such valuable information and for their permission to use their material.

We would also like to thank Ron Dunselman for making his paper available as well as Annette Jorry from Demeter International who translated some French texts into German.

Further thanks go to Georg Meissner and Werner Michtlits for their advice and support on this project.

We would particularly like to thank Jean-Paul Zusslin and Jean Baltenweck, as well as many wine growers and other friends, for making their photographs available.

Thanks are also due to Ambra Sedlmayer and Hans-Christian Zehnter for their patience and commitment with regard to the organisation and editorial work.

Such a project also needs financial support and we would like to warmly thank the Hamasil-Stiftung which provided support from the very beginning of the project as well as to the biodynamic wine producers who also provided financial assistance: Lalou Bize-Leroy, Domaine Leroy and Domaine d'Auvenay, Werner Michtlits, Domäne Meinklang, Marie-Thérèse Cappaz, Jean-Pierre Fleury, Domaine Fleury and Gérard Bertrand.

PICTURE CREDITS

aphotoflora.com: Figure 66

baumschule-horstmann.de: Figure 62

Bérillon, Lilian: Figures 13, 14, 15, 18, 19, 20, 21, 33, 34, 35, 36, 37, 79, 80, 81, 82, 83, 84, 85 and 88

Bockemühl, Jochen, *A Guide to Understanding Healing Plants, Vol 1 and 2*: Figures 1 and 61

Bournérias, Marcel; Pomerol, Charles; Turquier, Yves, *Le Golfe de Gascogne*: Figure 16

Campos, Antonio Luis: pp. 6, 8, 12, 35, 36, 68, 82 and 94

Colquhoun, Margaret and Ewalt, Axel, *New Eyes for Plants*: Figures, 5, 6 and 7.

commons.wikimedia.org: Figures 63 and 71

Dal, François, *Manuel des pratiques viticole contre les maladies du bois, Sancerre, Anger 2013*. Figures 72, 73, 74, 75, 76, 77 and 78

Fabian, Thomas: Figures 48 and 49

Florin, Jean-Michel: Figures 8, 9, 10, 17, 23, 24, 28, 39, 40, 59, 60, 67, 68, 69 and 91

Fournioux, Jean-Claude and Adrian, Marielle, *Morphologie et anatomie de la vigne*. Bordeaux 2011: Figures 25 and 26

Frits, Julius H, *Grundlagen einer phänomenologischen Chemie*. Stuttgart 1965: Figure 70

Geene, Jerome: Figure 57; pp. 2, 17, 65, 66, 84, 86, 104, 116, 138, 140, 188, 195 and 196

Hallé, Francie, *Plaidoyer pour l'arbre*: Figure 12

Hempel, Jörg: Figure 90; p. 136

Institut für Ökologie und Klimafarming: Figure 46

Izzo, Robert: Figures 50 and 51

Lemon, Ted: Figures 52, 53 and 54

loverofcreatingflavours.co.uk/2013/10/house-wine-organic-wine: Figure 39

Maack, Tom: Figure 58

MABD (Mouvement de l'Agriculture Bio-Dynamique): Figures 86 and 87

Meinklang: Figures 41, 42 and 43

Meyer, Frank, *Hepatordoron – ein veilseitges 'Typenmittel' der Anthroposophischen Medizin*. Der Merkurstab Nr. 3/2007, S.244–248: Figure 92

Michlits, Werner: Figure 44

National Forschungs- und Gedenkstätten der klassischen deutschen Literatur in Weimer: Figure 4.

Piollot, Jeanne: Figures 45 and 93

Rispens, Jan Albert, *Bäume versten lernen*: Figure 11

Searthwine.files.wordpress.com: Figure 55 (*marlboroughnz.com/listings/seresin-estate/*)

selbstsorger.de: Figures 64 and 65

Troll, Wilhelm, *Vergleichende Morphologie der höheren Pflanzen* [Comparative morphology of higher plants] Berlin 1967: Figure 29

Troll, Wilhelm, *Praktische Einführung in die Pflanzenmorphologie* [Practical introduction to plant morphology], Königsten 1975: Figures 30 and 31

Vinography.com: Figure 56

Zillig, Sylvia: Figures 2 and 3

Zusslin, Jean-Paul: Figures 32 and 47

CONTRIBUTORS

Daphne Amory

Works as an advisor and coach with land-based organisations in North America that are involved in biodynamic regenerative processes. Using tools and cues from the theatre of nature, she helps to realise the greater potential that is collectively unfolding.

Nikolaus Bolliger

Worked on various biodynamic farms, and since 1985 has run his own farm with emphasis on apple breeding. In 2004 he founded Poma Culta, a non-profit organisation that promotes research into biodynamic fruit breeding (*www.pomaculta.ch*).

François Dal

Since 2002 he has led the SICAVAC (Service Interprofessionnel de Conseil Agronomique, de Vinification et d'Analyses du Centre) laboratory for viticulture in Sancerre, France. He is engaged in research, consultancy and training.

Jean-Michel Florin

Born 1961 in Rosheim, France, Jean-Michel studied nature conservation and natural science at the Goetheanum in Dornach, Switzerland, with his main focus on landscape, medicinal plants and environmental education. Since 1988 he has been coordinator of the French biodynamic movement (Mouvement d'Agriculture Bio-Dynamique) and since 2010 co-leader of the Biodynamic Section at the Goetheanum.

Marc Follmer

Born in Strasbourg, France in 1955, Marc has a degree in pharmacy. For 20 years he has been a member of the management team of Weleda, France. He conducts research into substances in the context of life and their medicinal properties.

Robert Izzo

Worked in education, organisational development, and winemaking. During the mid-2000s Rob travelled extensively, visiting many respected wine regions. In 2012, with a growing interest in winemaking, Rob moved to Sonoma County, California, USA. He works as general manager at Eco Terreno, a winery committed to purpose over profit.

Ted Lemon

The first American ever to be selected as winemaker/vineyard manager of a Burgundian estate: Domaine Guy Roulot. In 1993 he and Heidi Lemon founded Littorai wines. By the late 1990s, Ted became convinced that western agronomic theory is both unsustainable and inadequate in its conception of plant health and nutrition. All of the vineyards that Littorai Wines owns or leases are farmed using biodynamic methods.

Georg Meissner

Studied oenology at the pharmaceutical faculty of the University of Montpellier, France. He has played a leading role in wine estates in South Africa and USA, and was engaged in humanitarian aid to Kosovo. He is currently the manager of Alois Lageder wine estate and is involved in training and research work at Geisenheim University, Germany.

Werner Michlits

Runs a multi-faceted Demeter farm with his family in Austria on the Hungarian border, with vineyard, orchard and livestock. In 2012 he founded the bilingual Waldorf school, Panonia, with his wife, with the aim of building a bridge between the two formerly united cultures. He also studied oenology in Geisenheim and was for many years active on the council of Demeter Austria and Organic Austria.

Colin Ross

Born in Malaysia in 1964, Colin grew up in Australia. As a young man he was a professional surfer, which brought him to Indonesia where he got to know its traditional agriculture. This led him back to Australia where he soon took on management of the Brookland Valley wine estate. During a visit to Burgundy he learnt about biodynamic agriculture and from 2006 to 2015 he managed the Seresin wine estate in New Zealand. He was also a member of the Biodynamic Council of New Zealand. He tragically went missing in 2017.

Matthias Wolff

Born in Freiburg in 1955, Matthias is a state recognised viticulture and wine technician. Since 1988 he has been consultant for organic wine and orchards in Baden-Württemberg. He is based at the Freiburg Viticultural Institute, and specialises in soil fertility and green cover crops.

Hans-Christian Zehnter

Born in Bochum, Germany in 1963, Hans-Christian studied biology and natural science at the Goetheanum in Dornach, Switzerland. From 2005 to 2010 he was editor of the weekly *Das Goetheanum*, and helped organise conferences at the Goetheanum. He is the author of books and articles and gives lectures and seminars focusing on the anthroposophical approach to nature observation (*www.sehenundschauen.ch*).

BIODYNAMIC ASSOCIATIONS

Demeter International
www.demeter.net

Australia:
Bio-Dynamic Research Institute
www.demeter.org.au

Biodynamic Agriculture Australia
www.biodynamics.net.au

Canada:
Society for Bio-Dynamic Farming
& Gardening in Ontario
www.biodynamics.on.ca

India:
Bio-Dynamic Association
of India (BDAI)
www.biodynamics.in

Ireland:
Biodynamic Agriculture
Association of Ireland
www.biodynamicagriculture.ie

New Zealand:
NZ Biodynamic Association
www.biodynamic.org.nz

South Africa:
Biodynamic Agricultural
Association of Southern Africa
www.bdaasa.ord.za

UK:
Biodynamic Association
www.biodynamic.org.uk

USA:
Biodynamic Association
www.biodynamics.com

NOTES

1. The Archetypal Plant

1   Goethe, *Scientific Writings,* Ch. 3 The experiment as mediator between object and subject.

2   Goethe, *Maxims and Reflections.*

3   Hermann Hesse, *Uber das Gluck: Gedichte und Betrachtungen.* Berlin 2013.

4   See Bockemühl, *Guide to Understanding Healing Plants,* Vol. 1.

5   This has been confirmed by the latest molecular genetic research. See for instance Fleur, *Pollination et fécondation.*

6   See Darwin, *The Power of Movement in Plants,* and Steiner, *World of the Senses,* lecture of January 1, 1912. Recent research appears to confirm this ever more strongly, where the root apex is referred to as a control centre that acts like a brain (S. Baluska, S. Mancuso, D. Volkmann, P. Barlow, Root apices as plant command centres – the unique 'brain-like' status of the root apex transition zone, *Biologia,* Bratislava, 59/Suppl. No. 13, 2004).

7   Schreiner, *Mycorrhizas and Mineral Acquisition in Grapevines.*

8   Hallé, Francis, *Plaidoyer pour l'arbre [A Plea for the Trees],* Arles, 2005.

9   Ibid.

10   Hallé adopted this hypothesis from the French chemist Gaudichaud de Beaupré (1789–1854).

2. The Vine

1   From Mathon, *Lucien Daniel, Inventeur du greffage créateur, fondateur de l'horticulture scientifique [Lucien Daniel, inventor of creative grafting, founder of scientific horticulture],* Association des Amis de Mitchourine, Paris 1955.

2   See also Miguel Altieri, 'The simplification of traditional vineyard-based agro-forests in north western Portugal, some ecological implications', www.researchgate.net/publication/226588164.

3   Bauer, *Grundgesten im Planzenreich.*

4   *Revue Horticole,* 1885, p. 220.

5   See Bauer, *Grundgesten im Pflanzenreich.*

6   The French botanist Francis Hallé confirmed this hypothesis by studying the sub-dividing tendency in other tropical species of *Vitaceae* (personal communication).

7   Ulrich, Roger, *La vie des fruits [The life of fruits]*, Paris, 1952.
8   Léglise, Max, *Les methodes biologiques appliquées à la vinification et à l'oenologie:vinifications and fermentations [Biological methods applied in the processing and making of wine: processing and fermenting]*, Paris, 2010.
9   See Steiner, *The Effects of Esoteric Development*, lecture of March 20, 1913.
10 See Steiner, *From Beetroot to Buddhism*, lecture of March 1, 1924, pp. 3f.
11 See Steiner, *Agriculture*, lectures of June 7 and 15, 1924.
12 Masson, *A Biodynamic Manual*, Ch. 18, pp. 273f.
13 Interview with Philippe Pacalet, *Le Monde*, Aug 28, 2013.
14 *www.adelaide.edu.au/wine-econ/databases/winegrapes*.
15 Pierre Masson, lecture at the winegrowers conference in Arbois, France 2016.
16 Jochen Bockemühl, 'Eine neue Sicht der Vererbungserscheinungen' [A new approach to inherited phenomena], *Elemente der Natutrwissenschaft*, No 100, 2014.

3. Pathogenesis: The Grape Louse Plague

1   Glöckler, *Waldorf Guide to Child Health*, p. 236.
2   Nick, *Schützen und Nützen: von der Erhaltung zur Anwendung, Fallbeispiel Europäische Wildrebe* [protect and utilise: from maintenance to utilisation, example of the wild vine] GenBank WEL. Hoppea, Regensburg Botanical Society, special edition 2014, pp. 159–73.

4. Salutogenesis: Restoring the Vine to Health

1   Steiner, *Bees*, lecture of December 10, 1923, pp. 89f.
2   Steiner, *Bees*, lecture of December 10, 1923, p. 90.
3   Steiner, *Agriculture*, lecture of June 14, 1924, p. 115.

5. The Agricultural Organism

1   Steiner, *Agriculture*, lecture of June 11, 1924, p. 184.
2   Steiner, *Agriculture*, lecture of June 10, 1924, p. 150.
3   See Oltmanns, *Vieharme Landwirtschaft. Brauchen wir Tiere für eine nachhaltige Bodenfruchtbarkeit? [Animal deficient landscape. Do we need animals to sustain soil fertility]*, Forschungsring Darmstadt, Materialien No. 27, 2013.
4   See Steiner, *Agriculture*, lecture of June 10, 1924, pp. 144f.
5   See Masson, *A Biodynamic Manual*, Ch. 17, p. 266.
6   Jean-Paul Zusslin, 'Ich bin ein Schaffer von Verbindungen' [I am a creator of connections], *Das Goetheanum* No. 12. p. 19.
7   See Bockemühl, *Waking up to Landscape*, and Colquhoun & Ewalt, *New Eyes for Plants*.
8   These few suggestions are intended to stimulate ideas. The Biodynamic Association in your country may offer courses in landscape design.
9   See Note 6 above.

8. Biodynamic Vineyards at Work

1   Purdy, *After Nature*, p. 237

9. Manure

1 Steiner, *Agriculture*, lecture of June 10, 1924, pp. 27f.

2 The research scientist Edwin Scheller coined the concept of active nutrient mobilisation through plant roots. See Scheller, *Grundzüge einer Pflanzenernährung*.

3 Steiner, *Agriculture*. lecture of June 10, 1924, p. 27.

4 Bockemühl, *Vom Leben des Komposthaufens [Life of compost heaps]*, Dornach, 1979.

5 Bockemühl and Järvinen, *Extraordinary Plant Qualities*.

6 Steiner, *Agriculture*, lecture of June 13, 1924, pp. 102f.

7 Steiner, *Agriculture*, lecture of June 13, 1924.

8 H. Spiess, C. Matthes, M. Hacker, 'Einfluss des Schafgarbenpräparates auf Kaliumentzug und Blattwachstum' [Influence of the yarrow preparation on potassium deficiency and leaf growth], *Lebendige Erde*, No. 1, 2000, pp. 34–36.

9 Steiner, *Agriculture*, lecture of June 13, 1924, pp. 99f.

10 There is a very beautiful description of the cow and its nature by Anet Spengler Neff in Hurter, *Agriculture for the Future*. See also Neff, Hurni and Steiff, *Why Cows Have Horns*.

11 For a more detailed description of how biodynamic preparations can be used see Masson, *A Biodynamic Manual*.

12 Paul Mäder, Andreas Fliessbach, David Dubois, Lucie Gunst; Padruot Fried, Urs Niggli, Fertility and Biodiversity in Organic Farming, *Science* No. 296, 2002, pp. 1694f.

10. Esca: a Grapevine Trunk Disease

1 Summary of Florian Bassini thesis, ENITA, Clermont Ferrand 2004. Article in *Biodynamis*, No. 50, 2005.

2 Henri Marès, *Memoire sur les maladies de la vigne en 1856* [Recalling the vine diseases of 1856].

11. Fungal Diseases

1 Francis Chaboussou, 'Physiologie et résistance de la plante' [The physiology and resilience of the plant], *Document Natur et Progrès*, No. 16, 1975, pp. 5–7.

2 Mathon, *Lucien Daniel*. (See Ch. 2, Note 1 above.)

3 B.A. Ferguson, et al. 'Coarse scale population structure of pathogenic Armillaria species in a mixed conifer forest in the Blue Mountains of north east Oregon', *Canadian Journal of Forest Research*, Vol. 33, No. 4, 2003.

4 Goethe, *Schriften zur Morphologie*, Vol. 1.3 p. 254.

5 Bockemühl & Järvinen, *Extraordinary Plant Qualities*, and Bockemühl, *Guide to Understanding Healing Plants*, Vol. 2.

6 Steiner, *Agriculture*, lecture of June 11, 1924, p. 60.

7 Strüh, Hans-Joachim, 'Equisetum und Kiesel' [Equisetum and silica], *Tycho de Brahe-Jahrbuch des Goetheanismus*, 1989, p. 167.

8 Steiner, *Agriculture*, lecture of June 14, 1924, p. 128.

9   Masson, *A Biodynamic Manual,* pp. 115–17, 275f.

10 Rispens, Jan-Albert, 'Weiden: Bäume am Wasser' [Willows: trees of water], *Merkurstab* No. 5, 2007, pp. 421–35.

11 Bockemühl, *Guide to Understanding Healing Plants,* Vol. 2.

12 For example between 182 mg in the Weleda garden in Arlesheim, Switzerland to nearly 1055 mg alongside the tram lines (also in Arlesheim); see also W. Daems, *Korrespondenzblätter für Ärzte,* No. 50, 1963, pp. 7–13.

13 Steiner, *Agriculture,* lecture of June 13, 1924, pp. 98f.

14 See Jochen Bockemühl, 'Entwicklungsbilder zur Charakterisierung von Löwenzahn und Brennessel' [Developmental characteristics of dandelion and stinging nettle], *Elemente der Naturwissenschaft,* No. 12, 1970, pp. 1–14.

15 Habermehl, Gerhard G; Hamman, Peter E; Krebs, Hans C, and Ternes, W, *Naturstoffchemie, eine Einführung [The chemistry of natural substances: an introduction],* Springer, Berlin, 2008, p. 530.

12. The Nature of Sulfur

1 Steiner, *Agriculture,* lecture of June 11, 1924, p. 54.

2 Steiner, *Agriculture,* lecture of June 11, 1924, p. 45.

13. The Nature of Copper

1 See, for instance, Pelikan, *The Secrets of Metals,* and Julius, *Fundamentals for a Phenomenological Study of Chemistry.*

2 See Etienne Laveau, Réduction des doses de cuivres en Viticulture Bio [reduction of copper doses in organic viticulture], Institut Technique de l'Agriculture Biologique 2009.

14. Pruning to Reflect the Vine's Nature

1 Steiner, *The Riddle of Man,* pp. 139f.

2 *https://simonitesirch.com.*

15. Gentle Pruning to Prevent Wood Disease

1 Drouhault, 'Greffage de la vigne en écusson' [vine grafting and budding], *Revue de Viticulture,* Vol. 4, No. 81, July 6, 1895, p. 11.

16. The Vine Nursery: Training the Breeder's Eye

1 Mathon, *Lucien Daniel.* (See Ch. 2, Note 1 above.)

19. Wine Gold

1 Goethe, *Theory of Colours,* preface, p. xvii.

2 Steiner, *Goethean Science,* Ch. 4, p. 45.

20. Developments in Apple Breeding: a Possibility for the Vine

1 Philippe Pacalet, 'L'homme qui veut semer des pépins de raisin' [the man who wanted to sow grape pips] *Le Monde,* Aug 26, 2013.

2   Basler, Pierre and Scherz, Robert, *PIWI Rebsorten [Fungus resistant vine varieties], Wädenswill, 2011.*

3   There are around 380,000 species of plant in the world according to the IUCN (International Union for Conservation of Nature and Natural Resources). See also Stefan Ungricht, 'How Many Plant Species are There?', *Taxon,* Vol 53, No. 2, May 2004, pp. 481–84.

4   See Timmermann, Martin, *Der Züchterblick, Erfahrung, Wissen und Entscheidung in der Getreidezüchtung [The breeder's eye: experience, knowledge and judgement in the breeding of cereals], Aachen, 2009.*

5   Steiner, *Agriculture,* lecture of June 7, 1924, pp. 20–25.

6   Steiner, *Agriculture,* lecture of June 10, 1924, pp. 34–36.

21. The Future of Wine: Working with a Creative Tension

1   Steiner, *From Beetroot to Buddhism,* lecture of March 1, 1924, pp. 3f.

FURTHER READING

Bockemühl, Jochen, *A Guide to Understanding Healing Plants,* Vol. 1, Mercury Press, USA, 2010.

—, —, Vol. 2, Mercury Press, USA, 2000.

—, *Waking up to Landscape,* Dornach, 1992.

— and Järvinen, Kari, *Extraordinary Plant Qualities for Biodynamics,* Floris Books, UK, 2006.

Colquhoun, Margaret and Ewalt, Axel, *New Eyes for Plants: Workbook for Plant Observation and Drawing,* Hawthorn Press, UK, 2002.

Darwin, Charles, *The Power of Movement in Plants,* New York, 1880.

Glöckler, Michaela; Goebel, Wolfgang, and Michael, Karin, *A Waldorf Guide to Children's Health,* Floris Books, UK, 2018.

Goethe, Johann Wolfgang von, *Maxims and Reflections*, Penguin Books, London, 1998.

—, *The Metamorphosis of Plants,* MIT Press, USA, 2009.

—, *Theory of colours,* (tr Charles Eastlake), CreateSpace Independent Publishing Platform, 2015.

Hurter, Ueli, *Agriculture for the Future,* Verlag am Goetheanum, Dornach, 2014.

Julius, Frits H. *Fundamentals for a Phenomenological Study of Chemistry,* AWSNA, USA, 2003.

Masson, Pierre, *A Biodynamic Manual,* Floris Books, UK, 2014.

Neff, Anet Spengler; Hurni, Beatrice, and Steiff, Ricco, *Why Cows Have Horns,* Forschungsinstitut für biologischen Landbau, (FiBL), Frick, Switzerland, 2016. Download available at www.fibl.org/en/shop-en/1712-cows-horns.html.

Pelikan, Wilhelm, *The Secrets of Metals,* Lindisfarne, USA, 2006.

Purdy, Jedediah, *After Nature: A Politics for the Anthropocene,* Harvard University Press, USA, 2015

Schreiner, R. Paul, *Mycorrhizas and Mineral Acquisition in Grapevines,* American Society for Enology and Viticulture, 2005.

Steiner, Rudolf. (Volume numbers refer to the Collected Works (CW).)

—, *Agriculture: Spiritual Foundations for the Renewal of Agriculture* (CW 327), Bio-Dynamic Association of North America, 1993 (also published as *Agriculture Course: The Birth of the Biodynamic Method,* Rudolf Steiner Press, UK, 2004).

—, *Bees* (CW 351), Anthroposophic Press, USA, 1998.

—, *The Effects of Esoteric Development* (CW145), Anthroposophic Press, USA, 1997.

—, *From Beetroot to Buddhism* (CW 353), Rudolf Steiner Press, UK, 1999.

—, *Goethean Science: Introduction to Goethe's Natural-Scientific Writings* (CW1), SteinerBooks, USA, 2018.

—, *The Riddle of Man* (CW 20), Mercury Press, USA, 1990.

—, *The World of the Senses and the World of the Spirit* (CW 134), Rudolf Steiner Press, UK, 2014.

INDEX

You may also be interested in...

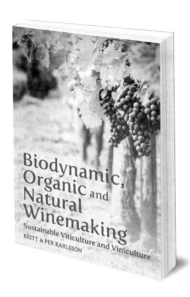

Biodynamic, Organic and Natural Winemaking

Sustainable Viticulture and Viniculture

Britt and Per Karlsson

'The combination of thorough research and personal interviews with growers and winemakers made this material come alive for me.'

– MIKE VESETH, THE WINE ECONOMIST

This comprehensive book by two renowned wine experts explains the rules, the do's and don't's of organic, biodynamic and natural wine production, both outside in the vineyard and in the wine cellar. It sets out clearly what a winemaker is allowed to do, including processes, additives and chemicals, and looks at the potential long-term benefits of going organic or biodynamic.

Winner of the 'Best Wine Book for Professionals' in the Gourmand International Cook Book Awards, Sweden.

florisbooks.co.uk

Have you tried our
When Wine Tastes Best App?

A fun and easy way to discover when wine will taste its best.

- Look ahead to key dates – search by month
- Swap between day and week view – see when wine tastes best at a glance
- Automatically adjusts to your time zone
- Try it for free – get a whole year's data with in-app purchase

Essential app for wine drinkers

More essential reading for biodynamic growers

The Biodynamic Orchard Book
Ehrenfried Pfeiffer and Michael Maltas

A Biodynamic Manual
Practical Instructions for Farmers and Gardeners
PIERRE MASSON

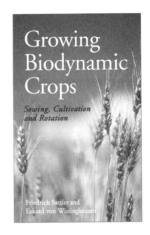

Growing Biodynamic Crops
Sowing, Cultivation and Rotation
Friedrich Sattler and Eckard von Wistinghausen

Pfeiffer's Introduction to Biodynamics
Ehrenfried Pfeiffer

Companion Plants and How to Use Them
Helen Philbrick & Richard B. Gregg

Koepf's Practical Biodynamics
Soil, Compost, Sprays and Food Quality   Herbert H. Koepf

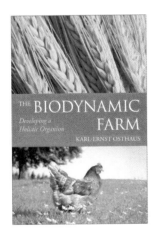

THE BIODYNAMIC FARM
Developing a Holistic Organism
KARL-ERNST OSTHAUS

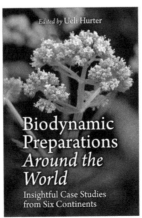

Biodynamic Preparations Around the World
Insightful Case Studies from Six Continents
Edited by Ueli Hurter

THE Maria Thun BIODYNAMIC CALENDAR
OVER 100,000 COPIES SOLD

florisbooks.co.uk

 Floris
Books

For news on all the latest books, and to get
exclusive discounts, join our mailing list at:

florisbooks.co.uk/mail/

And get a FREE book
with every online order!

We will never pass your details to anyone else.

Many wine growers have converted to biodynamic viniculture, often to combat the overuse of chemicals and help vulnerable vines. The quality of biodynamic wine is increasingly recognised around the world.

In this fascinating book, Jean-Michel Florin, coordinator of the French biodynamic movement, offers practical advice for both biodynamic wine growers and anyone considering converting to biodynamic methods. Beautifully illustrated with colour photographs throughout, this unique resource explores the nature of the vine, reveals the conditions required for healthy vine cultivation and looks to the future of vine research.

This in-depth exploration of biodynamic wine growing includes:

- an introduction to the Goethean method of observation in relation to vines;
- the theory of biodynamic viniculture;
- accessible and detailed articles on many aspects of wine growing, including biodiversity, pruning, treating and preventing disease;
- case studies of biodynamic vineyards from around the world, from France to the USA and New Zealand.

Drawing on contributions from worldwide biodynamic viniculture experts, this presents a positive global vision for the future of vine cultivation.

Floris Books

ISBN 978-178250-669-0

9 781782 506690   53500

£20.00
US $35.00

florisbooks.co.uk